RESEARCH IN GEOGRAPHY

RESEARCH IN GEOGRAPHY

Trends & Techniques

Anand Ballabh

AKANSHA PUBLISHING HOUSE
NEW DELHI - 110 002 (INDIA)

AKANSHA PUBLISHING HOUSE
4649/21-B, Ansari Road
Daryaganj, New Delhi - 110 002
Phones: 23263193, 9811582579
E-mail: ektabooks@yahoo.com

Research in Geography

First Published, 2007

ISBN 81-8370-076-4

PRINTED IN INDIA

Published by M.P. Misra for Akansha Publishing House, New Delhi - 110 002 and Printed at Capital Offset Press, Delhi

Contents

Preface

> He that does good for good's sake seeks neither praise nor reward, but he is sure of both in the end.
>
> —*William Penn*

Geography is the study of the Earth and its features and of the distribution of life on the earth, including human life and the effects of human activity. Geography research addresses both the questions of *where*, as well as *why* phenomena occur in particular places. The word comes from the Greek words Ge or Gaea, both meaning "Earth", and graphein meaning "to describe" or "to write".

Since place matters for everything from economics and health, to climate, plants and animals, geography is highly interdisciplinary. Geography draws upon and contributes to numerous other knowledge disciplines. Geography is broadly divided into human geography and physical geography, with subfields of geography including economic geography, political geography, urban geography, biogeography, geomorphology, coastal geography, and others.

Geography often seems to be involved in a process of questioning its own existence and subject—matter. Its boundaries are ill-defined and the discipline shares common concerns with other subjects--geology, history, economics, anthropology, sociology, planning, botany, politics, demography, and so on. Thus, geography cannot be easily located in the camp of either the earth sciences or the social sciences or, for that matter, the geometric sciences. The same is true when we examine the specialisms within geography. For instance, there is not an exact correspondence between geography and economics which produces economic geography. Economic geography also involves consideration of past treads; government intervention, and workforce characteristics and so has links with history, politics and planning, and demography. The difficulties of defining the area of study are even more acute in social geography, a rather young specialism, whose links with other disciplines are still being forged.

With advances in computer technology, the analytical and spatial data management tools available to geographers, including Remote Sensing, Geographic Information Systems and spatial data analysis, are now allowing geographers far more rigorous, quantitative analyses of spatial phenomena. Though, the qualitative approach to geography remains important, with ethnography and other methodologies used to investigate theories of spatial phenomena.

This book is organised under 38 chapters, encompassing all vital areas of geography. In the present book an endeavour is made to give a clear picture of geography and its branches and sub-fields. The chapters contained herein will give a brief description of major sub-fields, which can be further researched by the scholars.

While devising this book authoritative information is drawn from various secondary sources, like—write-up available on Net, books, abstracts, monographs, research papers, etc., penned by different scholars and geographers. Admittedly, the credit for this ought go to the original contributers.

We express our gratitude to all these geographers, the

men of great erudition, whose works or opinions are reporduced here. We are grateful to our friends, colleagues, librarians, teachers and well wishers for their guidance, supervision, moral support, inspiration and assitance while working on this project.

We will feel rewarded if this could help the readers in their academic pursuit. Suggestions and constructive criticism from all quarters are welcome to inhance the utility and quality of information.

1

What is Geography?

> "Mere place names are not geography. To know by heart a whole gazeteer full of them would not, in itself, constitute anyone a geographer. Geography has higher aims than this... to trace out the great laws of nature and to mark their influence upon man. In a word, geography is a science, a thing not of mere names, but of argument and reason, of cause and effect."
>
> —*William Hughes (King's College, 1863)*

I

Geography is the study of the Earth and its features and of the distribution of life on the earth, including human life and the effects of human activity. Geography research addresses both the questions of *where*, as well as *why* phenomena occur in particular places. The word comes from the Greek words Ge or Gaea, both meaning "Earth", and graphein meaning "to describe" or "to write".

Since place matters for everything from economics and health, to climate, plants and animals, geography is highly interdisciplinary. Geography draws upon and contributes to numerous other knowledge disciplines. Geography is broadly divided into human geography and physical geography, with subfields of geography including economic geography, political geography, urban geography, biogeography, geomorphology, coastal geography, and others.

With advances in computer technology, the analytical and spatial data management tools available to geographers, including Geographic Information Systems and spatial data analysis, are now allowing geographers far more rigorous, quantitative analyses of spatial phenomena. Though, the qualitative approach to geography remains important, with ethnography and other methodologies used to investigate theories of spatial phenomena.

II

HISTORY OF GEOGRAPHY

Ancient Greek Geography

The Greeks are the first known culture to actively explore geography as a science and philosophy. While maps had been made millennia earlier in Sumeria, the Greeks were the first to attempt to fully and systematically describe the world.

The ancient Greeks saw the poet Homer as the founder of geography. His works the *Iliad* and the *Odyssey* are works of literature, but both contain a great deal of geographical information. Homer describes a circular world ringed by a single massive ocean. The works show that the Greeks by the 8th century BCE had considerable knowledge of the geography of the eastern Mediterranean. The poems contain a large number of place names and descriptions, but for many of these it is uncertain what real location, if it exists, is actually being referred to.

Thales of Miletus is one of the first known philosophers

known to have wondered about the shape of the world. He proposed that the world was based on water, and that all things grew out of it. He also laid down many of the astronomical and mathematical rules that would allow geography to be studied scientifically. His successor Anaximander is the first person known to have attempted to create a scale map of the known world and to have introduced the gnomon to Ancient Greece.

Hecataeus initiated a different form of geography, avoiding the mathematical calculations of Thales and Anaximander he learnt about the world by gathering previous works and speaking to the sailors who came through the busy port of Miletus. From these accounts he wrote a detailed prose account of what was known of the world. A similar work, and one that mostly survives today, is Herodotus' *Histories*. While primarily a work of history, the book contains a wealth of geographic descriptions covering much of the known world. Egypt, Scythia, Persia, and Asia Minor are all described in great detail. Little is known about areas further a field, and descriptions of areas such as India are almost wholly fanciful. Herodotus also made important observations about geography. He is the first to have noted the process by which large rivers, such as the Nile, build up deltas, and is also the first recorded as observing that winds tend to blow from colder regions to warmer ones.

Pythagoras was perhaps the first to propose a spherical world, arguing that the sphere was the most perfect form. This idea was embraced by Plato and Aristotle presented empirical evidence to verify this. He noted that the Earth's shadow during an eclipse is curved, and also that stars increase in height as one moves north. Eudoxus of Cnidus used the idea of a sphere to explain how the sun created differing climatic zones based on latitude. This led the Greeks to believe in a division of the world into five regions. At each of the poles was an uncharitably cold region. While extrapolating from the heat of the Sahara it was deduced that the area around the equator was unbearably hot. Between these extreme regions both the northern and southern hemispheres had a temperate belt suitable for human habitation.

These theories clashed with the evidence of explorers, however. Hanno the Navigator had traveled as far south as Sierra Leone, and it is possible other Phoenicians has circumnavigated Africa. In the fourth century BCE the Greek explorer Pytheas traveled through northwest Europe, and circled the British Isles. He found that the region was considerably more habitable than theory expected, but his discoveries were largely dismissed as fanciful by his contemporaries because of this. Conquerors also carried out exploration, for example, Caesar's invasions of Britain and Germany, expeditions/invasions sent by Augustus to Arabia Felix and Ethiopia (Res Gestae 26), and perhaps the greatest Ancient Greek explorer of all, Alexander the Great, who deliberately set out to learn more about the east through his military expeditions and so took a large number of geographers and writers with his army who recorded their observations as they moved east.

The ancient Greeks divided the world into three continents, Europe, Asia, and Libya (Africa). The Hellespont formed the border between Europe and Asia. The border between Asia and Libya was generally considered to be the Nile river, but some geographers, such as Herodotus objected to this. Herodotus argued that there was no difference between the people on the east and west sides of the Nile, and that the Red Sea was a better border. The relatively narrow habitable band was considered to run from the Atlantic Ocean in the west to an unknown sea somewhere east of India in the east. The southern portion of Africa was unknown, as was the northern portion of Europe and Asia, so it was believed that they were circled by a sea. These areas were generally considered uninhabitable.

The size of the Earth was an important question to the Ancient Greeks. Eratosthenes attempted to calculate its circuference by measuring the angle of the sun at two different locations. While his numbers were problematic, most of the errors cancelled themselves out and he got quite an accurate figure. Since the distance from the Atlantic to India was roughly known, this raised the important question of what was in the vast region east of Asia and to the west of Europe. Crates of Mallus proposed that there were in fact four inhabitable land masses, two in each hemisphere. In Rome a large globe was created depicting this world. That some of the figures Eratosthenes had used in his calculation were considerably in error became known, and Posidonius set out to get a more accurate measurement. This number actually was considerably smaller than the real one, but it became accepted that the eastern part of Asia was not a huge distance from Europe.

While the works of almost all earlier geographers have been lost, many of them are partially known through quotations found in Strabo. Strabo's seventeen volume work of geography is almost completely extant, and is one of the most important sources of information on classical geography. Strabo accepted the narrow band of habitation theory, and rejected the accounts of Hanno and Pytheas as fables. None of Strabo's maps survive, but his detailed descriptions give a clear picture of the status of geographical knowledge of the time. A century after Strabo Ptolemy launched a similar undertaking. By this time the Roman Empire had expanded through much of Europe, and previously unknown areas such as the British Isles had been explored. The Silk Road was also in operation, and for the first time knowledge of the far east began to be known. Ptolemy's *Geographia* opens with a theoretical discussion about the nature and techniques of geographical inquiry, and then moves to detailed descriptions of much the known world. Ptolemy lists a huge number of cities, tribes, and sites and places them in the world. It is uncertain what Ptolemy's names correspond to in the modern world, and a vast amount of scholarship has gone into trying to link Ptolemaic descriptions to know locations.

Pliny the Elder's Natural History also has sections on geography.

For the most part Ancient Greek geography was an academic field. There is little evidence that maps or charts were used for navigation. It does, however, seem that at least in Athens the people were acquainted with maps and that several were on public display. It was the Romans who made far more extensive practical use of geography and maps.

Middle Ages

During the Middle Ages, geographical knowledge in Europe regressed (though it is a popular misconception that they thought the world was flat), and the simple T and O map became the standard depiction of the world. In the Middle East, however, Arabs such as al-Idrisi, Ibn Battuta, and Ibn Khaldun maintained the Greek and Roman techniques and developed new ones. The Islamic empire stretched from Morocco to India, and Arab traders travelled throughout Asia, Africa, and the Indian Ocean. The Arabs added a great deal of knowledge to expand and correct the classical sources.

Marco Polo to the 18th Century

Following the journeys of Marco Polo, interest in geography spread throughout Europe. The great voyages of exploration in 16th and 17th centuries revived a desire for both accurate geographic detail, and more solid theoretical foundations. The *Geographia Generalis* by Bernhardus Varenius and Gerardus Mercator's world map are prime examples of the new breed of scientific geography.

By the 18th century, geography had become recognized as a discrete discipline and became part of a typical university curriculum in Europe (especially Paris and Berlin), although not the in the United Kingdom where geography was generally taught as a sub-discipline of other subjects.

19th Century

One of the great works of this time was *Kosmos: a sketch of a physical description of the Universe*, by Alexander von Humboldt, the first volume of which was published in 1845. Such was the power of this work that Dr Mary Somerville, of Cambridge University intended to scrap publication of her own *Physical Geography* on reading *Kosmos*. Von Humboldt himself persuaded her to publish (after the publisher sent him a copy).

In 1877, Huxley published his Physiography with the philosophy of universality presented as an integrated approach in the study of the natural environment. The philosophy of universality in geography was not a new one but can be seen as evolving from the works of Alexander Von Humboldt and Immanuel Kant. The publication of Huxley physiography presented a new form of geography that analysed and classifed cause and effect at the micro-level and then applied these to the macro-scale (due to the view that the micro was part of the macro and thus an understanding of all the micro-scales was need to understand the macro level). This approach emphasised the empirical collection of data over the theoretical. The same approach was also used by Halford John Mackinder in 1887. However, the integration of the Geosphere, Atmosphere and Biosphere under physiography was soon over taken by Davisian geomorphology.

Over the past two centuries the quantity of knowledge and the number of tools has exploded. There are strong links between geography and the sciences of geology and botany, as well as economics, sociology and demographics.

The Royal Geographical Society was founded in England in 1830, although the United Kingdom did not get its first full Chair of geography until 1917. The first real geographical intellect to emerge in United Kingdom geography was Halford John Mackinder, appointed reader at Oxford University in 1887.

The National Geographic Society was founded in the USA in 1888 and began publication of the *National Geographic* magazine which became and continues to be a great popularizer of geographic information. The society has long supported geographic research and education.

20th Century

In the West during the second half of 19th century and the 20th century, the discipline of geography went through four major phases: environmental determinism, regional geography, the quantitative revolution, and critical geography.

Environmental Determinism

Environmental determinism is the theory that a peoples physical, mental and moral habits are directly due to the influence of their natural environment. Prominent environmental determinists included Carl Ritter, Ellen Churchill Semple, and Ellsworth Huntington. Popular hypotheses included "heat makes inhabitants of the tropics lazy" and "frequent changes in barometric pressure make inhabitants of temperate latitudes more intellectually agile." Environmental determinist geographers attempted to make the study of such influences scientific. Around the 1930s, this school of thought was widely repudiated as lacking any basis and being prone to (often bigoted) generalizations. Environmental determinism remains an embarrassment to many contemporary geographers, and leads to skepticism among many of them of claims of environmental influence on culture (such as the theories of Jared Diamond).

Regional Geography

Regional geography represented a reaffirmation that the proper topic of geography was study of places (regions). Regional geographers focused on the collection of descriptive information about places, as well as the proper methods for dividing the earth up into regions. Well known names from these period are Alfred Hettner from Germany and Vidal de la Blache from France. The philosophical basis of this field in United States was laid out by Richard Hartshorne who defined geography as a study of areal differentiation which later led to critic of this approach as overly descriptive and unscientific.

The Quantitative Revolution

The quantitative revolution was geography's attempt to redefine itself as a science, in the wake of the revival of interest in science following the launch of Sputnik. Quantitative revolutionaries, often referred to as "space cadets," declared that the purpose of geography was to test general laws about the spatial arrangement of phenomena. They adopted the philosophy of positivism from the natural sciences and turned to mathematics--especially statistics--as a way of proving hypotheses. The quantitative revolution laid the groundwork for the development of geographic information systems.

Critical Geography

Though positivist and post-positivist approaches remain important in geography, *critical geography* arose as a critique of positivism. The first strain of critical geography to emerge was humanist geography. Drawing on the philosophies of existentialism and phenomenology, humanist geographers (such as Yi-Fu Tuan) focused on people's sense of, and relationship with, places. More influential was Marxist geography, which applied the social theories of Karl Marx and his followers to geographic phenomena. David Harvey and Richard Peet are well-known Marxist geographers. Feminist geography is, as the name suggests, the use of ideas from feminism in geographic contexts. The most recent strain of critical geography is postmodernist geography, which employs the ideas of postmodernist and poststructuralist theorists to explore the social construction of spatial relations.

III

BRANCHES OF GEOGRAPHY

- Branches of physical geography
    - Geomorphology
    - Coastal/Marine studies

- Hydrology/Hydrography
  - Glaciology
  - Limnology
  - Oceanography/Oceanology
- Climatology
- Pedology
- Biogeography
- Geodesy
- Palaeogeography and paleoclimatology
- Environmental Geography and management
- Landscape ecology

- Branches of human geography
  - Economic geography
  - Transportation geography
  - Development geography
  - Population geography or Demography
  - Urban geography
  - Social geography
  - Behavioral geography
  - Cultural geography
  - Political geography, including Geopolitics
  - Historical geography
  - Time geography
  - Regional geography
  - Tourism geography
  - Strategic geography
  - Military geography

  - Feminist geography
  - Environmental geography
  - Children's geographies
- Geographic multi-disciplinary fields:
  - Cartography
  - Ethnography
  - Geographic information systems (GIS)
  - Geostatistics

General Topics

- Climate
- Demographics
- Earth Spheres
  - Atmosphere
  - Biosphere
  - Hydrosphere
  - Lithosphere
  - Anthrosphere
  - Cryosphere (Sometimes included in the hydrosphere)
  - Geosphere
  - Pedosphere (Soil)
- Geography
- Great Plains
- Landform
- Map
- Population density
- Population explosion

The Continents

- Africa
- Antarctica
- Australia
- Eurasia
- North America
- South America

Natural Geographical Features (Landforms)

- Bay
- Badlands
- Bluff
- Canyon
- Cape
- Cave
- Channel
- Cliff
- Coast
- Continent
- Coral reef
- Cove
- Crater
- Cuesta
- Dale
- Dell
- Delta
- Desert
- Escarpment

- Estuary
- Fen
- Fjord
- Foothills
- Forest
- Geyser
- Glacier
- Glen
- Gulf
- Gully
- Grassland
- Gully
- Highland
- Hill
- Hoodoo (geology)
- Inlet
- Island
- Knoll
- Lagoon
- Lake
- Landform
- Marsh
- Mesa
- Mountain
- Mountain pass
- Mountain range
- Ocean

- Plain
- Plateau
- Pond
- Prairie
- Peninsula
- Reef
- Ravine
- Ridge
- River
- Sandbar
- Savanna
- Sea
- Sound
- Source (river or stream)
- Spring (hydrosphere)
- Steppe
- Strait
- Stream
- Swamp
- Tide pool
- Tundra
- Vale
- Valley
- Vernal pool
- Volcano
- Waterfall
- Watershed

- Waterway
- Wetland

Manmade Geographical Features

- Airport
- Aqueduct
- Bridge
- Canal
- City
- Country
- Dam
- Farm
- Highway
- Megalopolis
- Nation
- National park
- Nature reserve
- Orchard
- Park
- Pipeline
- Ranch
- Reservoir
- Road
- Tunnel
- Tree farm
- Viaduct
- Wildlife reserve

The Greeks are the first known culture to actively explore geography as a science and philosophy. Mapping by the Romans as they explored new lands added new techniques. During the Middle Ages, Arabs such as Idrisi, Ibn Batutta, and Ibn Khaldun maintained the Greek and Roman techniques and developed new ones.

Following the journeys of Marco Polo, interest in geography spread throughout Europe. The great voyages of exploration in the 16th and 17th centuries revived a desire for both accurate geographic detail, and more solid theoretical foundations. This period is also known as Great Geographical Discoveries. By the 18th century, geography had become recognized as a discrete discipline and became part of a typical university curriculum in Europe (especially Paris and Berlin).

Over the past two centuries the quantity of knowledge and the number of tools exploded. There are strong links between geography and the sciences of geology and botany, as well as economics, sociology and demographics. In the West during the 20th century, the discipline of geography went through four major phases: environmental determinism, regional geography, the quantitative revolution, and critical geography.

Physical Geography

Physical geography (or physiogeography) focuses on geography as an Earth science. It aims to understand the physical features of the Earth, its lithosphere, hydrosphere, atmosphere, pedosphere and global flora and fauna patterns (biosphere). Physical Geography can be divided into the following broad categories:

- Geomorphology

- Coastal/Marine studies
- Hydrology/Hydrography
    - Glaciology
    - Limnology
    - Oceanography/Oceanology
- Climatology
- Pedology
- Biogeography
- Geodesy
- Palaeogeography and paleoclimatology
- Environmental Geography and management
- Landscape ecology

Human Geography

Human geography is a branch of geography that focuses on the study of patterns and processes that shape human interaction with various environments. It encompasses human, political, cultural, social, and economic aspects. While the major focus of human geography is not the physical landscape of the Earth, it is hardly possible to discuss human geography without referring to the physical landscape on which human activities are being played out, and environmental geography is emerging as a link between the two.

Human geography can be divided into many broad categories, such as:

- Economic geography
- Development geography
- Population geography or Demography
- Urban geography

- Social geography
- Behavioral geography
- Cultural geography
- Political geography, including Geopolitics
- Historical geography
- Time geography
- Regional geography
- Tourism geography
- Strategic geography
- Military geography
- Feminist geography
- Environmental geography
- Children's geographies

Socio-environmental Geography

During the time of environmental determinism, geography was defined not as the study of spatial relationships, but as the study of how humans and the natural environment interact. Though environmental determinism has lost support, there remains a strong tradition of geographers addressing the relationships between people and nature. There are two main subfields of socio-environmental geography:

- *Cultural and Political Ecology (CAPE):* Cultural ecology grew out of the work of Carl Sauer in geography and a similar school of thought in anthropology. It examined how human societies adapt themselves to the natural environment. Sustainability science has been one important outgrowth of this tradition. Political ecology arose when some geographers used aspects of critical geography to look at relations of power and how they affect people's use of the environment. For example, an influential study by Michael Watts argued that famines in the Sahel are caused by the changes in the region's political and economic system as a result of colonialism and the spread of capitalism.

- *Risk-hazards research:* Research on hazards began with the work of geographer Gilbert F. White, who sought to understand why people live in disaster-prone floodplains. Since then, the hazards field has expanded to become a multidisciplinary field examining both natural hazards (such as earthquakes) as well as technological hazards (such as nuclear reactor meltdowns). Geographers studying hazards are interested in both the dynamics of the hazard event and how people and societies deal with it.

Related Fields

Urban, Regional and Spatial Planning

Urban planning, regional planning and spatial planning use the science of geography to assist in determining how to develop (or not develop) the land to meet particular criteria, such as safety, beauty, economic opportunities, the preservation of the built or natural heritage, and so on. The planning of towns, cities and rural areas may be seen as applied geography although it also draws heavily upon engineering, the arts, the sciences, lessons of history, and politics. Some of the issues facing planning are considered briefly under the headings of rural exodus, urban exodus and smart growth.

Regional Science

In the 1950s the regional science movement led by Walter Isard arose, to provide a more quantitative and analytical base to geographical questions, in contrast to the descriptive tendencies of traditional geography programs. Regional science comprises the body of knowledge in which the spatial dimension plays a fundamental role, such as regional economics, resource management, location theory, urban and regional planning, transport and communication, human geography, population distribution, landscape ecology, and environmental quality.

Geographic Techniques

As spatial interrelationships are key to this synoptic science, maps are a key tool. Classical cartography has been joined by a more modern approach to geographical analysis, computer-based geographic information systems (GIS).

- *Cartography* studies the representation of the Earth's surface with abstract symbols (map making). Although other subdisciplines of geography rely on maps for presenting their analyses, the actual making of maps is abstract enough to be regarded separately. Cartography has grown from a collection of drafting techniques into an actual science. Cartographers must learn cognitive psychology and ergonomics to understand which symbols convey information about the Earth most effectively, and behavioral psychology to induce the readers of their maps to act on the information. They must learn geodesy and fairly advanced mathematics to understand how the shape of the Earth affects the distortion of map symbols projected onto a flat surface for viewing. It can be said, without much controversy, that cartography is the seed from which the larger field of geography grew. Most geographers will cite a childhood fascination with maps as an early sign they would end up in the field.

- *Geographic information systems (GIS)* deal with the storage of information about the Earth for automatic retrieval by a computer, in an accurate manner appropriate to the information's purpose. In addition to all of the other subdisciplines of geography, GIS specialists must understand computer science and database systems. GIS has revolutionized the field of cartography; nearly all mapmaking is now done with the assistance of some form of GIS software.
- "Geostatistics" deal with quantitative data analysis, specifically the application of statistical methodology to the exploration of geographic phenomena. Geostatistics is used extensively in a variety of fields including: hydrology, geology, petroleum exploration, weather analysis, urban planning, logistics, and epidimeology. The mathematical basis for geostatistics derives from cluster analysis, discriminant analysis, and non-parametric statistical tests, and a variety of other subjects. Applications of geostatistics rely heavily on Geographic Information Systems, particularly for the interpolation (estimate) of unmeasured points.
- *Geographic qualitative methods*, or ethnographical; research techniques, are used by human geographers. In cultural geography there is a tradition of employing qualitative research techniques also used in anthropology and sociology. Participant observation and in-depth interviews provide human geographers with qualitative data.

In their study, geographers use four interrelated approaches:

- Systematic—Groups geographical knowledge into categories that can be explored globally.
- Regional—Examines systematic relationships between categories for a specific region or location on the planet.
- Descriptive—Simply specifies the locations of features and populations.
- Analytical—Asks *why* we find features and populations in a specific geographic area.

In the present book an endeavour is made to give a clear picture of geography and its branches and sub-fields. The following chapters will give a brief description of major sub-fields, which can be further researched by the scholars.

2

Environmental Geography

Nature is a revelation of God; art a revelation of man.

—*Henry Wadsworth Longfellow*

1797. New opinions are always suspected, and usually opposed, without any other reason, but because they are not already common.

—*John Locke,*
An Essay Concerning Human Understanding

I

Environmental geography is the branch of geography that describes the spatial aspects of interactions between humans and the natural world. It requires an understanding of the dynamics of geology, meteorology, hydrology, biogeography, and geomorphology, as well as the ways in which human societies conceptualize the environment.

The links between cultural and physical geography were once more readily apparent than they are today. As human experience of the world is increasingly mediated by technology, the relationships have often become obscured.

Environmental geography represents a critically important

set of analytical tools for assessing the impact of human presence on the environment by measuring the result of human activity on natural landforms and cycles.

II

ENVIRONMENTAL GEOGRAPHY AND NATURE: A HISTORICAL ACCOUNT

No exaggeration, 'Nature' is a nebulous word different with connotation. If nature is usually taken as referring to the world 'out there'-from the smallest grasshopper, through the Grand Canyon, to the most distant galaxy—it is also believed that there is a force at work, that nature is working according to certain principles and that if we study nature we can deduce a moral lesson.

Nature cannot simply be regarded as what is out there—a physical universe which preceded the world of human values, and which will presumably outlive the human race-because what is out there keeps changing its meaning. Every attempt at describing nature, every value attributed to Nature—harmonious, ruthless, purposeful, random—brings nature inside human society and its values.

Remarkably Western culture refers to a singular force called 'Nature' which is frequently personified as a woman: Dame Nature in mediaeval mystery plays; Mother Nature in everyday speech. Raymond Williams has pointed out that in other cultures, nature is represented as a complex network of forces: a spirit of the rain, a spirit in the wind, and so on. The Western model of nature-singular and often personified-is connected with the fact that we live in a culture based on monotheism. And in mediaeval culture. Nature was described as God's deputy-responsible for carrying out His wishes on Earth. One may speculate that this relationship embodied a compromise between Christianity and an older, animistic way of interpreting the world: a way of seeing the Earth as a living body, of seeing live forces in rocks and winds as well as in trees and animals. Although this animistic vision was superseded by the Christian world view and then by scientific values, it is extremely tenacious and clings on to us through figures of everyday speech: even today it is common to describe the way in which farmers allow fields to lie fallow for a year as "giving the soil a chance to rest".

Nature was not a process 'out there' to be analysed or exploited. Within the Scholastic world view (the mediaeval synthesis of Aristotelian and Christian doctrines), nature was a complex chain connecting God to the humblest pebble; and man was emphatically inside the system: above the other creatures by virtue of his reason, but below the angels. This belief system was not just about nature as we understand it today; it also embraced the structure of society. In Renaissance maps, it was conventional to represent the wind as gusts of breath which issued from the mouths of cherubs and angels. This was but one element of the way in which the Earth was perceived as a living body: the circulation of water through rivers and seas was comparable to the circulation of blood; the circulation of air through wind was the breath of the planet; volcanoes and geysers were seen as corresponding to the Earth's digestive system-eruptions were like belches and farts issuing out of a central stomach.

Descriptions of the state talked of the "body politik" in which the princes were the rulers—the brain—and the Church was the soul. The distribution of wealth was the job of the financial system, just as the stomach and intestines were responsible for the distribution of food. And the feet upon which society rested were the peasants and artisans. It was a very hierarchical model, but one which stressed the interdependence of all the elements. Correspondences not only existed between our bodily function and the social order; they also existed between the human body and the universe: this is the view that underlies astrology. Each part of the body was thought to be governed by a zodiacal sign, so that the whole body was a miniature replica (or microcosm) of the celestial spheres.

Furthermore, the personality of each man and woman was said to be described by the positions of the stars at the moment of birth, so that we all carry within ourselves a model of the universe. Through this system of correspondences, man: society: the universe were unified into a coherent system. A system that was replaced by the rise of a scientific outlook which broke the individual, society, and nature down into their component parts to explain how the bits fitted together.

Natural Resources

The change from one outlook to another was not a clean break; for some two hundred years these two systems overlapped. There was confusion about what was alive and what wasn't. When Harvey gave the first clear account of the circulation of the blood (1628), many geologists were excited: they thought this was the way in which water circulated the planet. If blood can flow uphill, then why not water? The Jesuit scholar, Athanasius Kircher, was one of the first to study volcanoes and geysers at close quarters. In the pictures that he drew of these phenomena, we can see the beginnings of a scientific study, but also the belief that water flowed around the world through a circulatory system; and his pictures of mountains conveyed his view that they could be the skeletons of gigantic animals.

This is the picturesque side of a change in world view, but there was also a moral and political dimension. If the world were thought of as alive, this placed some restraints on how the world was treated. We are familiar with this idea from other cultures; we know that American Indians regarded the Earth as a great mother and found the white man's way of exploiting the Earth abhorrent:

> We trace out all the veins of the Earth and yet are astonished that it should occasionally cleave asunder or tremble: as though these signs could be any other than expressions of the indignation felt by our sacred parent!

Pliny also suggests that the results of mining are of dubious benefit to humanity: that gold has contributed to human corruption and avarice, while iron has led to robbery and warfare. Such restraints against the exploitation of resources were still active until the Renaissance, but they were dismantled as commercial mining gathered momentum during the fifteenth century. In an allegory which was published in Germany in 1495, this conflict between respect for the Earth and the commercial interests of mining was dramatised in the form of a vision. A hermit falls asleep and dreams that he witnesses a confrontation between a miner and Mother Earth-"noble and freeborn, clad in a green robe, who walked like a person rather mature in years". Her clothing is torn and her body has been pierced. She is accompanied by several gods who accuse the miner of murder. Bacchus complains that his vines have been fed into furnaces; Ceres states that her fields have been devastated by pollution; and Pluto says he cannot reside in his own kingdom because of all the hammering going on. In his own defence, the miner argues that Earth "who takes the name of mother and proclaims her love for mankind" in reality conceals metals in her inward parts in such a way that she fulfils the role of a step-mother rather than a true parent. Since gold turned into money is the surest way to create wealth, the mining of metal will help the poor, will decorate the churches as an act of piety, and will advance culture through the building of schools and the payment of teachers.

A further significant step in this process is to be found in the work of Francis Bacon (1561-1626) who is celebrated as the founding father of scientific research, and the innovator of inductive reasoning. Bacon was also a politician: Lord Chancellor of England under James I, who had written a book on witchcraft (*Daemonologie,* 1597). In 1603, the first year of his English reign, James passed a law condemning all practitioners of witchcraft to death. Carolyn Merchant suggests a striking congruity between the scientist's interest in nature, and the state's interest in witchcraft. Bacon proposes an experimental methodology for investigating nature, using language which is starkly sexual in its metaphors and suggestive of a witch-finder in its techniques. Advocating a scientific method which will be experimentally verifiable, displaying consistent results when repeated. Bacon writes: "For you have but to follow, and as it were hound Nature in her wanderings, and you will be able to lead her and drive her to the same place again." Describing the investigation of nature in the laboratory. Bacon invokes the language of the torture chamber: "I mean in this great plea to examine Nature herself and the arts upon interrogatories. For like as a man's disposition is never well known till he be crossed, nor Proteus ever changed shape till he was straitened and held fast, so Nature exhibits herself more clearly under the trials and vexations of art mechanical devices) than when left to herself."

Feminist historians have argued that the persecution of witches in the sixteenth and seventeenth centuries can be understood only as part of the revolution in values that took place as capitalism superseded the feudal system. A new order came to predominate in the seventeenth century which can be characterised as male, mathematical and mercantile. Male surgeons replaced female midwives in Britain, and women who practised medicine were likely to find themselves stigmatised as witches. Mercantile capitalism viewed nature as a resource to be exploited, as can be seen from the example of mining. And scientific research produced a mathematical model of the universe, which superseded the organic analogies of the animate Earth.

Copernicus (1543) attributed to the Earth a daily rotation on its axis, as well as an annual orbit around the sun, because this was a more mathematically concise system; Kepler, equipped with the data of Tycho Brahe, conceived a universe which was constructed from harmonious geometric relations.

Galileo (1638) succeeded in establishing mathematical principles which predicted the paths of both heavenly bodies and projectiles on Earth. For Galileo, "Natural philosophy is written in that great book which lies before our eyes—I mean the Universe-but we cannot understand it if we do not first learn the language and grasp the symbols in which it is written. The book of nature is written in the language of mathematics." Galileo followed Kepler in differentiating between primary and secondary qualities, but in a more pronounced form. Primary qualities included number, motion, and weight. These qualities were absolute, objective, and above all could be quantified mathematically. Secondary qualities included colour, taste, odour, and these qualities were relative, and above all subjective. These three scientists, through their observation of the heavens and their measurements of phenomena on Earth, drafted a mathematical model of the universe which amounted to a re-definition of nature. The real world lay outside of men and women—it was mathematics and astronomy and the study of motion. Objective nature was whatever could be measured.

Finally, Sir Isaac Newton stood on the shoulders of the scientists who had preceded him and reduced the major phenomena of the universe to a single mathematical law-the universal law of gravitation (1687). "The fundamental task of all science is to explain all phenomena in terms of matter and motion", proclaimed Newton. This view of the universe not only altered the status of man, it also made God slightly redundant. In the mediaeval world picture, God had been the "ultimate good" and man's purpose was to know God and to love Him. In Newton's system. God had been relegated to the role of laboratory technician whose job was to prevent the fixed stars falling out of the heavens, and to mend the defects of the celestial clockwork. Newton's favourite proof of the existence of God was the concentric rotation of all the planets in the same plane. The "ultimate good" was now the cosmic order of masses in motion, and man's role was to applaud the workings of this clockwork universe.

The Long March: From Mathematical to Pastoral Nature

Changes in the philosophy of nature are reflected in the human versions of nature that kings and gardeners construct. Perhaps the best gardening metaphor for the mathematical model of nature bequeathed to us by the seventeenth century is to be found in the garden constructed at Versailles for Louis XIV. From 1661 until 1690, under the direction of the king's gardener Le Notre, thousands of workers struggled to construct a magnificent formal garden out of a swamp. (There was a high death rate due to malaria.) Versailles became the capital of France and Louis XIV's statement of his absolute power.

Exactly, one hundred years later, a reaction against this account of nature as absolute order was sweeping across Europe. One can glimpse this at Versailles: hidden away in a corner of the king's estate is a fairy tale village constructed for Marie Antoinette, wife of Louis XVI. It is called Petit Hameau (Small Hamlet) and includes a mill, a farmhouse, and a dairy grouped around a lake. This village formed a backdrop for acting out a fantasy version of rural life: the queen and her servants would dress up as milkmaids and shepherds and frolic on the grass. This was the degenerate end of a view of nature known as the pastoral.

Pastoral poetry forms a tradition within Western culture that stretches back two thousand years to the verses of Virgil and Theocritus. The pastoral celebrated nature as restorative: an antidote to the sophistication and cynicism that characterised life at court. Later poets (such as Milton in *Lycidas)* used the pastoral as a way of attacking corruption in the church and in moral matters. They were not attempting a realistic description of country life, but engaging in a moral critique. The pastoral can be viewed as the seed from which sprang the Romantic view of nature: nature as embodying moral values. Nature as a corrective to man's pride. But by the eighteenth century, the pastoral had become the self-indulgence of Marie Antoinette's village, the amorous make-believe to be found in the paintings of Boucher and Fragonard: a fantasy of rural life untouched by work or poverty.

The pastoral was degenerating into a mood of Arcadian escapism at the very moment when fanners and landowners were beginning to systematically 'improve' the performance of nature. The advent of seed-drilling, the introduction of new crops and a more scientific approach to the breeding of livestock were all features of early eighteenth-century agriculture. If the pastoral were to retain any plausibility, it had to find a way of engaging with the realities of management and work within the eighteenth-century landscape.

The English artist Thomas Gainsborough toned down the artificiality of pastoral painting and drew on more realistic accounts of country life that were being painted in Holland by Hobemma and van Ruysdael. But to reconcile the detailed materialism of the Dutch style with the idyllic fantasies of French artists, such as Watteau and Poussin, produced tensions when it came to representing work.

A sense of the picturesque came to dominate the landowning classes' view of their property. Instead of nature's being marshalled into a grand design (as at Versailles), human mastery over it had become subtler, more softly modulated, more naturalised. From the increased wealth generated by commerce and industry, a rural retreat could be constructed as an escape from the smoke *of* the city. Through technical breakthroughs in drainage and earthmoving, it was how possible to re-arrange the landscape to suit the owner's taste. As Raymond Williams describes it, the English landscape garden was a *tour de force* of stage design, but: fully realised in physical terms: "a rural landscape emptied of rural labour and of labourers: a sylvan and watery prospect with a hundred analogies in neo-pastoral painting and poetry, from which the facts of production had been banished".

A classic example of this form of gardening can be seen at Stourhead in Wiltshire, designed by the owner. Henry Hoare II, a banker, and constructed under his direction between 1740 and 1780. This garden cannot be seen from the house, and was planned rather as the setting for walks and rides within a picturesque and literary landscape. Hoare had the valley to the west of his house dammed, creating a lovely series of lakes amidst wooded slopes. And a sequence of temples and mythological sites suggest literary associations to the visitor making his way round this carefully constructed "natural" landscape.

Two Natures: Industrial and Romantic

By the end of the eighteenth century, the industrial revolution was drawing the poor out of the countryside and propelling them towards the factories and cities. Around the year 1840, Britain became the first country where the ; urban population exceeded the rural population. The violence of this social transformation had a strong effect on the values that attached to the country and the city. More and more, the countryside was perceived as that which had been lost, that which had been left behind. From this perspective of social upheaval, William Wordsworth articulated the moral values that resided in nature.

Nature had become something we communed with, something we yearned for. Emphatically separate from human society and everyday life. It's worth pausing to discuss what it was that alienated and mystified nature into this fleeting, transcendental force. The short answer would be industrialisation and city life. As the centre of gravity of society was drawn towards the vast and fast-growing cities of the nineteenth century, the countryside and nature were felt to be repositories of a lost way of life, and of values that were in danger of extinction. A rural community was felt to be organic and rooted in the soil; in contrast to urban life with its anonymity and sense of constant change. Obviously the idea persists that closeness to the flow of nature—the smell of damp earth, the cyclical growth and decay we associate with the seasons-is somehow a more 'natural' way of life. And yet the industrial world is simply based upon another model of nature which enables the resources of the world to be analysed and broken down and re-shaped by human hands.

Francis Bacon had pointed this out in 1623, when he described three forms of nature: She is either free and follows her ordinary course of development as in the heavens, in the animal and vegetable creation and in the general way of the universe; or she is driven out of her ordinary course by the perverseness, insolence, and forwardness of matter, as in the case of monsters, or lastly, she is put in constraint, molded, and made as it were new by art and the hand of man, as in things artificial."

Today, a common meaning for nature is simply an escape from the "man"-made world: an Eden which we look forward to re-gaining at the weekend, or on holiday. It is a common experience to walk through a. beautiful, wooded landscape, and then to notice a line of electricity pylons and reflect how human industry has spoilt nature once again. It should be remembered that the electricity pylons, the very notion of an electric current as a controlled flow of electrons, are based upon an analysis and re-arrangement of nature.

From a particulars sense, the scientific version of nature gave rise to the Romantic version: to supply what was missing. Since sensations and emotions had been banished by science from nature, dismissed as subjective, the

Romantics expressed their alienation by constructing an account of nature that overflowed with emotions and moral feelings. They tried to replace the precision of science with a sense of mystery and infinity.

One of the most spectacular examples of changing accounts of nature reflecting changes in society can be seen in nineteenth-century England. Since the discoveries of Newton, the character of nature had been that of a constitutional lawyer: interpreting and classifying examples, making predictions out of precedents. "What was classically observed was a fixed state, or fixed laws, or fixed laws of motion. The laws of nature were indeed constitutional, but unlike real constitutions they had no effective history. What changed this emphasis was of course the evidence and the idea of evolution: natural forms had not only a constitution but also a history." At the beginning of the nineteenth century, the role of nature shifted towards the selective breeder, improving the species through natural selection. Perhaps the first significant figure in this process was Thomas Malthus, whose *Essay on Population* (1798) argued that the population was growing too quickly, and that finite resources of food would soon be exhausted. The replenishing cycles of nature would be wiped out by uncontrollable growth. Therefore such catastrophes as plague and famine were not aberrations, but an important part of nature's design. There were not enough places at nature's dinner table to feed everyone, suggested Malthus. Famine provided the pruning shears to thin out nature's garden. Harvest failures were examples of nature's red pencil auditing the book of life. The dinner table, the gardening shears, the audit book: all these images of a prudent, commercial society were used to legitimise a profound callousness towards 'natural' catastrophes.

Darwin's account of evolution. *The Origin of Species* (1859), was remarkable for the controversy it provoked about whether mankind was part of nature, as well as for the more ferocious character that was attributed to nature as a consequence of Darwinism. When Bishop Wilberforce, a mathematician and ambitious prelate, debated with Thomas Huxley, one of the first scientists to declare his support for Darwin's theories, Wilberforce asked Huxley, "was it through his grandfather or his grandmother that he claimed his descent from a monkey?" The activities of Biblical fundamentalists in the United States who, in the 1970s, have tried to replace the theory of evolution with the Book of Genesis, may remind us that this question is not yet dead. But though the theory of evolution has become a part of scientific orthodoxy, perhaps Darwin's work has had a still more profound (if unintended) effect on ideas of human society.

Artists as well as poets dramatically revised the personality of nature, moving from the pastoral to the violent. The most celebrated nature painter of mid-nineteenth-century Britain, Sir Edwin Landseer, painted landscapes of natural carnage. *The Swannery Invaded by Eagles* depicts a group of snow-white swans being torn to shreds by sea eagles. *Man Proposes, God Disposes* features two thuggish polar bears munching the bones of some hapless polar explorers. And an early work by Landseer, *The Cat's Paw,* illustrates an ingenious act of animal sadism: a monkey seizes a cat's paw and forces it against a red-hot stove. Many of his contemporaries were disturbed by Landseer's skill at representing animal savagery, and it is difficult to look through the body of his work without feeling that these images of animal violence are really statements about society.

The way in which Darwin's argument influenced accounts of society is obvious from such everyday figures of speech as the rat race, the pecking order, the law of the jungle. As soon as Darwin's theory had become popular currency, historians and social scientists saw history in a new light.

Karl Marx, at work on *Das Kapital,* acknowledged that Darwin's theory of natural selection could serve "as a basis in the natural sciences for the class struggle in history". Marx wrote to Darwin, offering to dedicate *Das Kapital* to the great scientist. But his collaborator, Friedrich Engels, warned that: "The distinguishing feature of human society is production and when the means of production are socially produced, the categories taken from the animal kingdom are inapplicable." Similarly, the anarchist Kropotkin pointed out that Darwin's argument concerned competition *between* species, whereas Social Darwinism seeks to justify competition *within* a species: between members of the human race. And yet there has been no shortage of analogies from Darwin to society because the idea seems so suited to competition for jobs on the labour market, competition between companies for a share of the market, competition between nations for territory and power. In short, the salient features of capitalist, industrially advanced, and politically rivalrous powers.

John D. Rockefeller argued that you cannot grow a perfect example of the American Beauty rose without first debudding it of minor blooms; in the same way, a company cannot achieve maturity without eliminating serious competitors-in other words, establishing a monopoly. The German commander von Moltke (who provided the military muscle behind Bismarck's unification of Germany) saw war as the supreme example of the Darwinian struggle for existence, where the armies of competing nations clashed and only the strongest survived. On a global level, it could even be argued that European nations had a moral duty to compete with 'inferior' races, in order to raise the quality of the species.

Europe during end of 19th century was in command of Africa and vast areas of Asia. The perspective on nature shifted from the state of all-out war that characterised the 'Darwinian' outlook, to the need for efficient management. Having conquered foreign territories, the task was to set up an efficient government. Having eliminated your industrial competitors, the job was now to consolidate your market. Having collected all the animals and plants, it was now necessary to catalogue them.

Major museum collections were systematised to give order to the specimens that had been slaughtered around the world. In Britain, Lord Rothschild despatched expeditions to remote corners of the globe to collect butterflies, and the Rothschild Collection ultimately amounted to two-and-a-half million set butterflies and moths, meticulously labelled and mounted, 300,000 birds, 144 giant tortoises, and 200, 000 birds' eggs. Some naturalists have speculated that this spirit of insatiable collecting and cataloguing is a consequence of the Protestant ethic at work in the animal kingdom, since the major European collections were to be found in Britain, Germany, and Sweden; in Catholic countries this tendency towards monumental collections of nature hardly existed.

The Royal Botanic Gardens at Kew housed a similar collection of plants uprooted from all over the world. Research and selective breeding of plants took place at Kew; important crops were made more 'productive' and then transplanted to other locations within the British Empire. In particular, the rubber plant was successfully 'kidnapped' from its home in the Amazon Basin, improved upon, and then transplanted to Malaya where it became the basis of a multi-million-pound industry.

Even in municipal parks and gardens, it became common to incorporate ducks and trees transported from other continents. Henry Hoare's garden at Stourhead was transformed by importing exotic rhododendra from North America and the Himalayas, so today the garden has a very different character from its original one. Nature as a global resource was being gathered together and efficiently cultivated.

Simultaneously, the warfare between town and country was mellowing into the idea that these two states could be harmoniously integrated. Howard's diagram for his "slumless, smokeless city" was a series of concentric circles with six boulevards radiating out from the centre, where the major civic buildings were located. Industrial, commercial, and residential zones were sited in different parts of the city and broken up by 'green belts'. The outermost circle was an agricultural zone which would supply the city with food, and also each house was to have an extensive garden to produce vegetables for the family. Through this scheme, Howard believed the most alienating effects of industrial capitalism could be transformed: instead of being divided by competition, the city would nourish through a spirit of co-operation. The polar opposites of town and country would be superseded by "a third alternative, in which all the advantages of the most energetic and active town life, with all the beauty and delight of the country, may be secured in perfect combination". The first garden cities were built at Letchworth, Hampstead, and Welwyn; these formed the starting point for a peculiarly British form of Utopian planning. Another generation of New Towns were built after the Second World War, and in the 1960s Milton Keynes was started as the largest (and probably last) city in this line of thinking. To some extent, the vision of transforming society was reduced to a series of town planning procedures; the Housing and Town Planning Acts of the twentieth century incorporated many of Ebenezer Howard's ideas, and the project of reconciling country and city became a bureaucratic system.

Rhaticus suggested that if you can measure something, then you have some control over it. This was an interesting metaphysical notion. The French mathematician Laplace took things one stage further when he postulated: "If a superhuman intelligence were acquainted with the position and motion of every atom in the universe, then this intelligence could predict the entire future of the universe." It was assumed that this mathematical hypothesis could never be put to the test. But with the growth of computer technology, and more and more sophisticated ways of gathering information, encoding it, processing it, and disseminating it, one can see Laplace's remark as more than just abstract speculation. The more the world is analysed and quantified, the more sophisticated the intelligence gathering and control systems become.

To answer the question, "What has all this to do with nature?" It could be argued that a world characterised by an explosion of information and information processing systems is the culmination of the project that began with the scientific revolution: to decode the book of nature as mathematical information. The essence of computer technology is that information can only be fed in once it has been encoded in digital form. Any information that cannot be mathematically encoded, cannot be considered.

In the modern context, the Newtonian world view is certainly in control of the planet, and this growth of information systems has amounted to a new industrial revolution. The first industrial revolution analysed nature in order to imitate it mechanically, and improve its efficiency. Steam engines were far more powerful and efficient than water wheels or windmills, the old ways of deriving power from sources of energy. The new industrial revolution is electronic rather than mechanical, and is concerned with encoding information and simulating aspects of the human nervous system. Machines that can count. Machines that can calculate. Machines that can read and write, that can talk and listen, that can see and analyse what they see, that can recognise voices and faces.

The more complex the imitation of the human brain, and the more it is linked to mechanical systems which can act on that information, the closer we come to a world which exists as a vast simulation of so-called reality.

The ways in which these systems amount to forms of political control are complex and quite subtle. When Rhaticus suggested that measurement is a form of control, he did not necessarily mean straightforward coercion. The earliest Greek scientist, Thales of Miletus, caused a sensation in 585 B.C. when his astronomical measurements enabled him to predict an eclipse of the sun. His mathematical knowledge was a primitive form of power. Knowledge itself brings control. In more down to earth terms, one can also notice how this electronic revolution is automating thousands of jobs out of existence. Word processors will replace many secretaries and clerks; And it is now common in any high street to programme some digits into an electronic cash dispenser, and be handed some bank notes by a machine. When computers are linked to mechanical systems, the result is the science of robotics which enables human workers to be mechanically replaced. The jobs that do remain can be paced and subjected to close surveillance.

Systems for monitoring movements and for recording and analysing phone conversations, telex messages, radio signals, are all in the forefront of this electronic revolution. An electronic imitation of the world might seem to be the antithesis of nature-but it derives from the way in which Galileo and Newton reinterpreted nature as a mathematical system. Today could be characterised as the cybernetic era of nature-nature as bits of information. This is the culmination of the doctrine of primary qualities.

Summing Up

Town planner, ruthless predator, selective breeder, "guardian of my moral being", constitutional lawyer, agricultural improver, celestial clock-maker, "wanton harlot", loving mother: all these characters have been attributed to nature in the past five hundred years. Which one do we live with today? The short answer would be all of them. When we watch a TV programme about nature, we are likely to be startled by close-ups of creatures tearing each other to shreds. If we switch to a gardening programme, we may see middle-aged men with moustaches speaking tenderly about a young plant they are nurturing, and touching its leaves as gently as a baby. If we visit the natural history museum, we can study nature stuffed and mounted and painstakingly catalogued. If we take a holiday in the Lake District or the Himalayas we may be overwhelmed by the grandeur of nature. If our holiday is interrupted by an earthquake, we are reminded of nature's monstrous, destructive forces. If we simply go for a walk in a park, we are likely to glimpse a version of either the mathematical order of Versailles, or of the carefully constructed landscape at Stourhead. And if we shop for natural shampoo or for herbal remedies, perhaps we are shopping for a reminder of the days when the Earth was alive.

III

ENVIRONMENTAL DETERMINISM

Environmental determinism, also known as *climatic determinism*, *environmentalism*, or the *geographic factor*, is the view that the physical environment, rather than social conditions, determines culture. Those who believe this view say that humans are strictly defined by stimulus-response (environment-behavior) and cannot deviate.

Carl Ritter, Ellen Churchill Semple, Ellsworth Huntington, and Thomas Griffith Taylor were prominent environmental determinists. Friedrich Ratzel was mistakenly viewed as deterministic by many of his students. Jared Diamond's work is often characterized as a revival of environmental determinism.

Environmental determinism's origins go back to antiquity, when the Greek geographer Strabo wrote that climate influences the psychological disposition of different races. Similar ideas continued to be propounded up into the modern era.

Environmental determinism rose to prominence in the late 1800s and early 1900s when it was taken up as a central theory by the discipline of geography (and to a lesser extent, anthropology). Clark University professor Ellen Churchill Semple is credited with bringing the theory to the United States after studying with Friedrich Ratzel in Germany. The prominence of determinism was influenced by the high profile of evolutionary biology, although it tended to resemble the now-discredited Lamarckism rather than Darwinism.

The basic argument of the environmental determinists was that aspects of physical geography, particularly climate, influenced the psychological temperament of individuals, which in turn shaped the culture and society that those individuals formed. For example, tropical climates were said to cause laziness and promiscuity, while the frequent weather changes of the middle latitudes led to sharpened intellects. Because these environmental influences operate slowly on human biology, it was important to trace the migrations of groups to see what environmental conditions they had evolved under.

In the 1920s and 1930s, environmental determinism came under attack as its claims were found to be hopelessly inadequate at best, and often destructively wrong. Geographers reacted first by developing the softer notion of "environmental possibilism," and later by abandoning the search for theory and causal explanation for many decades. Later critics charged that determinism served to justify racism and imperialism. The experience of environmental determinism has left a scar on geography, with many geographers reacting negatively to any suggestion of environmental influences on human society.

A variant of environmental determinism was popular among Marxists at the same time. To Marx's basic model of the ideological and cultural superstructure being determined by the economic base, they added the idea that the economic base is determined by environmental conditions. For example, Russian geographer Georgi Plekhanov argued that the reason his nation was still in the feudal era, rather than having progressed to capitalism and becoming ripe for the revolution into communism, was that the wide plains of Russia allowed class conflicts to be easily diffused. This Marxist environmental determinism was repudiated around the same time as classic environmental determinism.

REFERENCES

Arthur Lovejoy, *The Great Chain of Being* (Harvard University Press, 1936).

Athanasius Kircher, *Mundus Subterraneus* (Amsterdam, 1665). See Marjorie Nicolson, *Mountain Gloom and Mountain Glory* (New York: Cornell University Press, 1959), pp. 168-73.

Bacon, 'De Dignitate et Augmentis Scientarum' in *Works,* ed. J. Spedding, R. Ellis. D. Heath (London: Longmans, 1870). vol. 4, p. 296.

Carolyn Merchant, *The Death of Nature* (London: Wildwood, 1982), 'Organic Society and Utopia', pp. 69-99.

Edwin Arthur Burtt, *The Metaphysical Foundations of Modern Physical Science (London:* Kegan Paul, 1925); ch. 3: 'Galileo', pp. 61-95.

John Barrell, *The Dark Side of the Landscape: The Rural Poor in English Painting 1730-1840* (Cambridge University Press, 1980).

Lucile Brockway, *Science and Colonial Expansion: The Role of the British Royal Botanic Gardens* (London: Academic Press, 1979), pp. 141-65.

Marjorie Nicolson, *Newton Demands the Muse* (Princeton University Press, 1946).

Merchant, *The Death of Nature;* Susan Griffin, *Woman and Nature* (New York: Harper and Row, 1978).

Miriam Rothschild, *Dear Lord Rothschild* (London: Hutchinson, 1983).

M.H. Abrams. *The Mirror and the Lamp* (Oxford University Press, 1953). pp. 298-335.

Niavis, Judicum Jovis (Leipzig, n.d.). For a fuller account of this allegory, see Frank Dawson Adams, *The Birth and Development of the Geological Sciences* (New York: Dover, 1938), pp. 171-5.

Pliny, *Natural History,* trans. J. Bostock and H. T. Riley (London: Bohn, 1858), vol. 6, Bk 33, ch. 1.

R.M. Young, 'The Human Limits of Nature', in J. Benthall (ed.). *The Limits of Human Nature (London:* Allen Lane, 1973).

R.M. Young, 'Malthus and the Evolutionists: The Common Context of Biological and Social Theory', *Past and Present,* 43 (1969), pp. 109-45.

R.M. Young, 'The Historiographic and Ideological Contexts of the Nineteenth Century Debates on Man's Place in Nature', in M. Teich and R. Young (eds.). *Changing Perspectives in the History of Science* (London: Heinemann, 1973), pp. 344-438.

Raymond Williams, *Keywords* (London: Fontana, 1976), pp. 184-9.

Raymond Williams, *Problems in Materialism and Culture* (London: Verso, 1980), 'Ideas of Nature', pp. 67-86.

Raymond Williams, *The Country and the City* (London: Chatto & Windus, 1973), ch. 12: 'Pleasing Prospects'.

Williams, *Problems in Materialism and Culture,* 'Social Darwinism'. pp. 86-103.

3

Social Geography and Geography of Society

In calling the 20th century the "era of the common man," we tend to forget that it took the genius of very uncommon men to make this modern world possible. For example, an ordinary citizen can today buy at a corner drugstore, for a relatively small sum, a wonder drug for which Louis XIV would have given half of France. By any standard, Edison, Marconi, Pasteur and Salk were more gifted than we ordinary mortals, and certainly we all enjoy the fruits of their genius. What man could weave his own suit or construct his own electric light? These conveniences, these unearned rewards are so much a part of our daily lives that it seems to me each citizen's daily prayer should be, "Forgive me, O Lord, for taking so much for granted."

—*James A. Farley*

I

Case Study: India

Victims of their own historical success, Indians suffer from a rural economy. The reason? A high population density, poverty and strong traditional caste system holds back any progress or urbanisation. The fertile Ganges Valley with monsoon rain and river always supported a dense rural population. Rice is the stable crop. A settled traditional agriculture is practised on small plots, but tenants are exploited by landlords. There is a large mass of landless labourers. Poverty still acute, however, the emerging middle class peasantry benefited from the Green Revolution.

Areas of Study

Questions in the field of social geography might include the examination of rural exodus or urban exodus or whether low-rise developments generate a different type of daily life than tower blocks. It deals also with problems of segregation and discrimination, socio-spatial variations in health, analysis of spatial crime patterns and others.

In the field of community development (or community economic development), the importance of place has been a focal point for sociologists to determine what effects geography may have on a local community's cohesiveness and the sense of community. Studies in community psychology suggest that *where* we are many times has an effect on *who* we are.

II

Space is not simply a surface upon which changes. Changes within the structure of economic production involve the geographical reorganisation of labour and capital; and, in turn, the changing geography of economic activity affects the shape and composition of the workforce and throws up new cultural patterns and political configurations. Changes in the technology of production are central to this process of geographical and social reorganization—changes which involve a rearrangement and control of natural forces to achieve profitable and competitive conditions of production. Gold captured this view of nature when he spoke of the electronic imitation of the human brain, microprocessor technology, or 'nature as bits of information' as he refers to it, as the latest objective quantifiable arrangement of nature. This way of approaching the relationship between society, space and nature, this kind of geography, alters the manner in which economic, political and cultural change is conceived. However, the implications of this view, are not merely conceptual, they also raise methodological questions.

Conceptual and methodological issues are woven together to produce a particular kind of geographical approach. The methods of analysis and synthesis employed in the following two sections do not in themselves produce a new kind of geography. The terms are familiar within the discipline of geography.

In fact, the two modes of approach, analysis and synthesis, are best understood as complementary methods that enable us to understand the geographical organization of society, both at a detailed level and at a richer, more comprehensive level. The process of analysis involves the selection and isolation of a particular aspect of society, the consideration in detail of its various features and the relationships that hold between them, including their geographical characteristics, at the expense of other aspects of society to which they are related. The value of this type of exercise rests upon a number of pragmatic grounds. First, we cannot hope to take up every aspect of society that has some bearing upon, say, the changing structure of Britain's economy, and simultaneously study all aspects with the same analytical intensity. In a social world structured by a series ofcomplex, changing interrelationships, the process of analysis allows us to hold certain relationships apart and to concentrate in detail upon each aspect in turn. This internal focus is the nub of analysis; it allows an investigation in depth of one aspect of a broader picture. Second, some kind of division and separation of the different aspects of the geography of the social world is necessary in order to enable us to address particular social issues and problems.

A geography of housing, of welfare, of employment, is the practical development of the need to outline handlable issues and formulate specific answers to specific questions. Which issues are defined as manageable, which questions are raised is not pre-given, and invariably rests upon the institutionalized sub-divisions that hold sway within a discipline at any one point in time.

Portrayed as such, there is nothing that is particularly new either to geographers or to other social scientists about the task analysis is expected to perform. What is distinctive about the selection of readings in the following section, however, is that the analyses of the three topics—cultural forms, urban economic activity, and the process of international law--are framed in terms of how their geographical organization affects the very way in which they work.

The geographical unevenness of local cultures, their unique characteristics, and challenges the thesis of a mass culture that pervades all areas and regions of the UK. From this standpoint he shows how the cultural differences between places have had a profound effect upon the shape and absorption of new cultural configurations. New cultural forms, he argues, are not simply mapped onto local cultures: the relationship between the two is one of negotiation and resistance, 'traces' of earlier local cultures affect the form of new cultural strands. In turn, he argues that the geography of culture cannot be understood without reference to structural changes in the occupational division of labour.

Changes in the structure of capitalist production which have reworked the social and gender composition of the workforce also, he points out, involve a geographical shift of capital and labour. The movement of both industry and workers to new locations has led to the disruption of old cultural boundaries and created a new pattern of cultural unevenness and flux. Cultural, economic and spatial change are, he argues, combined in the overall process of social change. The links between economic, political and cultural processes are built into his analysis, but they are deliberately played down to allow a more intensive focus upon changing cultural traditions. In an analytical sense, Clarke has turned a spotlight upon cultural processes to show how culture itself cannot be fully understood outside of its geographical context. Ball and Picciotto draw out the same point in relation to urban economic activity and international law respectively. Ball demonstrates how the cost of crossing space measured in terms of time, information and money influences and shapes the way in which different types of economic activity, both private and public, use urban space.

The barriers of distance may be used, for example, by retailers to gain a monopoly position over a particular clientele by locating at a distance from their competitors' shops. In this sense, space is integral to the activities of retailers and cannot be analysed in isolation from their economic behaviour, nor can their economic behaviour be considered in isolation from spatial opportunities and spatial form. At the level of the city as a whole the spatial form of social processes can produce contradictory results. As the tertiary and quaternary sectors take over the city centre they push up land prices there, and their interest in land and location as a long-term asset pushes them up even more. One result is that low-income housing becomes, given the politics and economics of the production of housing, less and less of a feasible proposition. And yet it is precisely the office-owners of the tertiary and quarternary sectors that demand, close at hand, the nightly army of low-income workers to clean up after the day's business.

Picciotto addresses the growing incidence of Jurisdictional conflicts between major capitalist nation states. To put it another way, the increasing internationalization of capital has geographically outstripped the political process of internationalization. Accordying to Political territoriality, is more than the geographical space occupied by a nation state, it is an assertion of power and control over the events and relationships that occur within a defined space. The anchorage of states to place which allows the defence of cconomic interests within a territorial unit is shown by Picciotto to be at odds with a process of capital that has outgrown state political boundaries.

Conflicts over territorial jurisdiction have arisen as the concomitant of the interpenetration of capital investment between the developed capitalist economies. The changing geographical structure of the international economy and the political world order are integral to the form in which the conflicts have arisen and are played out.

Ball's analysis of urban economic life does not exhaust all that there is to know about the dynamic forces which shape and structure city life. For the purposes of detailed investigation he selects, or abstracts, three elements--the division of labour, population distribution and the built environment—from, say, the political forces and cultural processes which also shape the fabric of city life.

In making such a choice Ball is consciously dividing

up and separating off different aspects of the geography not only of the city, but also of the wider social world of which cities are a part. But he is also careful to draw the links, for example, between an urban spatial division of labour and the wider processes of capital accumulation and national state policy. His analysis spills over into social relationships that appear only at the edge of his focus, yet indicate the direction in which changes within cities may be understood in a more holistic fashion. This is the groundwork for the construction of a broader synthesis of which the city would be one element. In a similar vein Picciotto's analysis of Jurisdictional conflicts between capitalist nation states indicates that an issue of international law cannot be adequately comprehended outside of its relation to aspects of international economics and politics. The latter two features are undeveloped in his account, the links are faintly drawn, to enable him to portray the issue ot legal conflict in sharp relief. Clarke, also insistent that culture should not be seen as simply a set of local differences in isolation from wider aspects of society. The complex geography of culture that he outlines is not conjured up by a variety of local characteristics; local cultural terms draw their changing shape from a combination of such characteristics and the wider changes that have affected the economic organisation of British society, both at a national and at an international level.

There is a mundane level at which everyone is familiar with a geography of British culture. To visit a museum, art gallery, theatre or cinema is to be confronted by the geography of cultural institutions. Trips to the town or city centre consume time and money. To live outside London involves a recognition of the concentration of cultural resources and institutions of all kinds which divides the 'metropolis' from the 'provinces'.

These patterns bear the mark of social processes. Here, the 'civic pride' of nineteenth-century philanthropists who stamped their mark on city centres with galleries and museums, and there the 'economic rationality' of the cinema chains, closing the old suburban 'flea-pits' and developing the city centre multiple-screen complexes. Space and place are essential elements in the patterns of British culture, and one of their most profound effects is to be found in the way they structure cultural diversity. Blackpool and Eastbourne offer different holiday pleasures, not solely because of where they are, but because of the different social groups whose holiday needs they cater for. The bank holidays and wakes weeks of a Northern working class, stretched for fifty weeks a year by labour and poverty, required an intensity of sensation and satisfaction garishly reflected in the distorting mirrors of Blackpool's 'cheap thrills'. The inter-war middle classes of the Southeast, by contrast, lived in a culture defined by the virtues of familiar decency and respectability more adequately serviced by the genteel spaces and tea rooms of the South Coast.

Importance of Place in Culture

In fact, place—which may be the region, the city and the neighbourhood—condenses a whole complex history of economic, social and political processes into a simple cultural image. The persistent opposition of North versus South expresses itself in cultural images (hard v. soft; rough *v.* cultured; straight *v.* sly) but such stereotypes are meaningless without attention to the economic, social and political differentiation which are condensed in these parodies. The cultural division refracts the economic and industrial unevenness of Britain, and the concentration of economic power and control in the Southeast.

It captures something of the distribution of 'cultural

capital' which flows alongside that concentration of economic capital, and it gestures knowingly at the focus of political power that sits just around the corner from the City. It is, of course, a parody. It subsumes class within region, as if there is neither Southern working class nor Northern capitalists. But it is a parody which contains a partial, and uncomfortable, recognition of the geography of economic power and control in Britain. The demarcation of Britain into distinctive cultural regions owed much to the way capitalism focused its development.

The North of England was dominated by the urban centres in which the heavy industrial sectors and textiles were concentrated. The Midlands, the Black Country and Birmingham in particular, were organized around metal and secondary engineering production, experiencing an influx of labour from Scotland and Wales. The Southeast, although dominated by the centre of financial and political power in London, was also becoming the favoured site for the development of the new light industries' in engineering, electrical and chemical industries. In the Northeast, Scotland and Wales, the older skilled trades, in sectors such as mining and shipbuilding, were bearing the brunt of the inter-war Depression.

This uneven distribution of industrial sectors explains something of the differential impact of the Depression on British society. The decline of the old 'staple' industries, and the unemployment and pauperization which followed, scarred the face of Northern England, Scotland and Wales in sharp contrast to the self-confident expansion experienced in the Southeast. British culture, the ways in which people experienced, understood and responded to these material conditions, bore the marks of this uneven and unequal economic structure. The different 'cultures' which were present in inter-war Britain reflected these disparate economic conditions through a variety of national, regional and local identities. So, the Welsh working class, shaped by chapel, rugby and Celtic cultural legacies, possessed a culture distinct from that of the Northern working class, based around football and cricket, the regional identities (and rivalries) of Yorkshire and Lancashire, and the 'civic pr de' of the mill and steel towns.

To deal with the full interplay of Britain's inter-war economic, social and cultural patterns here is not possible. Each 'place' was culiurally complex. Scotland contained the diversity of agricultural highlands, the Glasgow working class, Edinburgh's 'cultural' cosmopolitanism, and so on. Similarly, within any town, different areas and neighbourhoods contained specific cultural patterns from the 'exclusive' residential districts to run-down terraced housing. But what should be clear is that place—nationally, regionally and locally--played a major role in organizing these cultural differences, and was an essential *symbol* in the identification of difference.

Geordie, Glaswegian, Mancunian, Cockney: each local identity condensed a whole range of economic, social and political references into a place. But this geography of culture is subject to drift, erosion and dramatic eruptions. The processes which forged the class cultures—and their places—of the inter-war years did not stop, leaving these cultures frozen into a static formation. Viewed from the 1980s, these pictures of Blackpool and Eastbourne become nostalgic snapshots of a different time, a different way of life now transformed by social change. The clear boundaries of social distinction have been washed away. Between then and now stands the Second World War, affluence, new technology, the expansion of the mass media--and recession. What, then, has happened to the old places,, boundaries and patterns, and, more pertinently, what has replaced them?

Mass Culture Thesis

The mass culture thesis pointed to the massive post-war expansion of the mass media (television) as the force which had abolished cultural difference. The spread of television, and its penetration into the home, were seen as the means through which a generalized culture was being dispersed and absorbed. And since the commercial drive behind the media was concerned for audience maximization, the programming which sought to create these audiences would be based on the' lowest common denominator' approach. These three elements--spread, penetration and content--form the core of the mass culture thesis.

(1) The spreading ownership of televisions allowed most of the nation to be linked together as a mass receiver of transmitted culture. This linking of a mass audience to a single message promised the obliteration ofintranationa! cultural differences, and even the disappearance of international cultural differences. The communications revolution promised to make the world 'a global village'-the abolition of geography itself-with American its common language.

(2) Tevision penetrated to the 'sacred hearth', the world appeared in the living-room corner. Its visual immediacy, allied to its domestic setting. was understood to provide the culture it presented with especially powerful impact.

(3) The content of mass culture threatened to extinguish traditional standards. Since most mass cultural theorists were, like most mass culture, American,

they spoke of mass culture displacing the values embodied in 'high' and 'popular' culture.

High culture—the arts—was prized for its engagement of the emotions and intellect in the quest for knowledge and ultimate human values. 'Popular' culture, while not possessing these same virtues, was at least tolerable in that it demonstrated the vitality and robustness of'the people'. But mass culture threatened to produce nothing but the 'cultural dupe'-the passive and indiscriminate sponge.

Briefly, the changing character of labour, full employment, new housing, the new way of life based on the telly, the fridge, the car and the glossy magazines—all have had their effect on our political strength. (Quoted in Hall *et al.,* 1978, p. 230)

Gaitskell's summary of the forces which had eroded Labourism were taken up by cultural analysts concerned with the disappearance not only of working class politics, but also of a distinctive working class culture beneath the rolling waves of affluence and mass culture. Studies such as Richard Hoggart's *The Uses of Literacy* (1959) are ambiguous. They are, as Critcher has argued, "a response to... the argument that the working class had ceased to exist". While they celebrate the distinctiveness of working class culture, they also lament its anticipated disappearance. Hoggart's study offers both a sensitive description of a distinctive way of life, and an attempt to analyse the of its destruction—those proceses which are "unbending the mainsprings of action".

The studies of this period are diverse in their identification of the main causes of change (affluence, mass culture, social policy in housing, etc.) but each of them offers its testimony to the disappearance of something distinctive. And each of them also recognizes that one of the changes is the remaking of the geography of class.

The working class 'community' dealt with in these studies marked the convergence of geography and culture. Whether in a London housing estate (Young and Willmott's study of Bethnal Green), a Northeast mining village or Hoggart's Leeds, what these studies celebrated was a specific geographical density of social relationships and a shared and distinctive way of life. The economic, political and cultural changes of post-war Britain were identified as destroying the strengths of these densely supportive local class cultures. In their place, the future offered the affluent, individualized, middle-class culture of mass society. While these cultural and community studies looked to the old working class estates, towns and villages for the distinctiveness of working class culture, other sociologists were looking elsewhere for the signs of the future.

Goldthorpe, Lockwood, Bechofer and Platt, in their mammoth study of *The Affluent Worker* (1969), shook the dust of the North from their feet and set off for the Southeast. There, in the mass production systems and high wage economy of the car industry, the future culture of the working class might be discerned. Consequently, in the 1960s it became Luton whose entrails were examined for signs of the new working class. The signs were ambiguous and have been argued over ever since. The 'affluent worker' was not becoming middle class (the victim of "embourgeoisification'), but nor was Luton the setting for the continuation of the old culture of working class community. Instead the ' new' working class seemed to show an instrumental or calculative approach to employment, trade unionism and politics. Social life--the culture of these workers--had lost the focus of communal sociability and solidarity of the 'community' and replaced it with a 'privatized' or family-centred lifestyle, in which 'place' played no significant part. The factory and the home--not Luton--were the only places which counted.

Youth and Cultural Diversity

Apparently, The post-war outbreak of 'youth culture' secmed to represent further evidence of this 'convergence' of lifestyles. Cultural difference seemed now to be based on age rather than class, 'Teenage culture' appeared to be a classless world of distinctive music, clothes, clubs and behaviour. For those searching anxiously for the portents of mass culture, the signs were hardly encouraging. Youth, like the affluent workers, represented a' vanguard'--the first generation exposed to classlessness, affluence and the mass media. And the behaviour of this new generation reinforced all the fears of the mass culture pessimists simmorality, declining standards, drugs, violence and an apparent Wrrencter to mindless consumption. Both T.V. channels now run regular programmes in which popular records are played to teenagers and judged. While the music is performed, the cameras linger savagely over the faces of the audience. What a bottomless chasm of vacuity they reveal. Huge faces, bloated with cheap confectionery and smeared with chain store make-up, the open, sagging mouths and glazed eyes, the hands mindlessly drumming in time to the music, the broken stiletto heels, the shoddy, stereotyped, 'with-it' clothes: here, apparently, is a portrait of a generation enslaved by a commercial machine. (Johnson, 1964). The obsessive attention to the' generation gap' as the basis of youth culture successfully prevented any recognition that youth was far from being classless. What was read as being a youth culture was in fact composed of a diversity of youth subcultures in which the structuring hand of class played a powerful role. What was held up as the image of a classless-mass culture—was its opposite—the re-emergence of distinctive lifestyles heavily determined by class. By the time that skinheads were beating up hippies, it was clear that--for them, at least--the class war was not over.

In the end, of course, the secret had to come out: class had not gone away after all. A growing body of research indicated the persistence of poverty, gross inequalities of income and life chances and the stubborn refusal of class to be magically forgotten. In part, the excitement and despair about the arrival of mass culture—and the variety of analyses it generated--had been based on a misunderstanding. The very distinctiveness of the inter-war class cultures had deceived the eyes and minds of the analysts. The end of traditional working class culture was understood as marking the end of the working class. The cultural forms of class were taken as being the reality of class.

Because the distinctive working class communities were no longer visible geographically, and because the cultures were no longer visible socially, the assumption was made that class itself must have disappeared. The error was to assume that the *culture* was inseparable from the economic and social basis of class. Little thought was given to the ways in which the basis of class might be refracted through new and changing cultural expressions and practices. Watching television, taking package tours abroad, having a car changed working class culture but did not mark the end of class as an economic and social division."

This mistaken identification of culture with class was intensified by the distinction between the 'new' and the 'traditional' working class. Instead of dealing with the historical process of how classes and cultures have changed and developed, this distinction reduces history to one abrupt change—from the old to the new. The idea of the 'traditional' stops us thinking about history as a process, and leaves history only as a frozen image to be contrasted with the present. But those cultures now described as 'traditional' were themselves the product of economic and social processes, and the responses of social groups to their circumstances.

This is not to suggest that really nothing has changed since the Second World War: that because 'deep down' the class relations of capitalism are still firmly in place, no attention needs to be paid to superficial changes. At a very abstract level, the fundamental dynamic of labour and capital still obtains. But the studies of class cultures of the 1950s do, however erroneously, point to significant changes in the composition of class and culture.

The economic, occupational and social composition of the working class has been reworked by both the dynamic of capital in its search for new products, new markets and new profits and by policies of the state (in housing and other welfare areas). Although all of these processes are interconnected, it is worth briefly separating out some of their impacts. The last forty years have seen a constant process of capital redeployment in Britain The expansion of the home market for consumer durables—Gaitskell's cars, fridges and tellys--provided the basis for the post-war expansion of light engineering and the car industry in sites largely away from the 'old' heavy industries of the North. Goldthorpe *et al.'s* mission to Luton to find the 'new' (and affluent) worker caught something of this changing geography of the working class. As the working class became mobile, so the attachments to familiar places became disconnected. New Towns and new housing estates replaced the old communities. Pulled by the promise of affluent lifestyles, and pushed by the changing geography of employment (with the assistance of government regional policies), the working class set out to establish themselves in new places.

But these changes, in the period of post-war expansion, did not becomi-the permanent basis of new social and cultural patterns. By the 1970s, it was becoming clear that the processes of boom and affluence that had produced the new working class were in decline. British manufacturing was entering a deep recession, and the domestic market was being increasingly serviced by cheaper imported goods.

The rise and fall of the car industry is only one (though perhaps the most spectacular) of the occupational shifts, but others are equally significant for the composition of class. The growth of the 'service sector' (both state and private); the decline of the traditional male 'skilled' working class; the expansion of increasingly proletarianized white-collar occupations in technical and administrative systems have played a part in reorganizing the economic physiognomy of class. They have also had profound effects on the sexual division of waged labour. One aspect of the search for profitability has been the attempt to find sources of 'green' (inexperienced and non-unionized) labour, which has (together with the expansion of state and commercial 'service') fuelled the increase of women's employment.

It is apt to note here that these changing occupational patterns and their uneasy rhythms of decline and expansion have further reshaped the social geography of Britain. The concentration of decline in the old manufacturing industries has been particularly hard on the industrial-urban patterns of the North, Midlands, Scotland and Wales. The major cities and towns have suffered economically and socially, and within each, the process of decline has been most intense in what has become known as the 'inner city'. Equally acute has been the experience of towns dependent on single industries. The ' rationalization' of pits under the NCB removed the economic basis of whole villages and towns. while British Steel's closures at places such as Corby devastated whole local economies. The rise of new industries has taken place on new sites: oil-based industries in the East of Scotland, electronics in the Southeast 'sun belt' and small manufacturing operations around the New Towns and the industrial 'parks' on the peripheries of old towns. The once stable connection between urban and industrial patterns in British capitalism has been broken--both people and work are moving out of the big cities.

Once more, the patterns are uneven: nationally, regionally and locally. Scotland experienced simultaneous recession (in shipbuilding and the car industry, for example) and expansion (through the North Sea oil fields). Towns given the benefit of industrial enterprise zones to encourage growth lost jobs from companies outside such zones. As the rhythms of expansion and contraction have grown quicker through the post-war period (and as the depths, of contraction have intensified), so the certainty of 'place' has been dismantled. Fluidity and mobility became the cornerstones of economic policy, and the ties of place have been loosened in their wake. Nor can it be said that mobility itself was in any way a guarantee of a new place in the order of things. Workers who moved from Glasgow to the New Town of Linwood to work in the new ilootes car plant at the end of the 1960s experienced several changes of ownership which presided over a declining labour-force before the plant was finally closed."

The economic basis and social geography of class have undergone profound changes. But class is not solely a matter of production—of the organization of work within capitalism. The dynamics of capitalism also affect cultural patterns through the systems of distribution and consumption. The old cultures of region and locality owed as much to the social patterns of home, family and neighbourhood as they did to the conditions of employment. And the post-war period has seen equally substantial changes at this level. The residential patterns of towns and cities have been reshaped through the expansion of home-ownership, particularly concentrated in the new peripheral estates and New Towns--moving the more affluent out of the older inner city areas. Increasingly, the inner areas have been left to council redevelopment or to owner-occupation of older housing. The rise of the 'property-owning democracy' has intensified the social division and distinctions between centre and periphery in urban areas.

This movement to the peripheries involves the social reorganization of family life—making private transport an essential rather than luxury item. Individual mobility—the ability to get to work, to the shops, to social and leisure activities—has become a necessity for social mobility. The car, while valued for the independence which it provides, has simultaneously created dependence. When public transport has been steadily reduced almost everywhere, those without access to a car have experienced growing social isolation. For the elderly, the unemployed, the young and a majority of women, the 'privatization' of British society, its increasing home-centredness. is as much an enforced condition as it is a freely chosen way of life.

The pressure towards mobility has been intensified by changes in the system of distribution of consumer goods and services. The extent of concentration and rationalization of capitalist production is well understood, but rather less attention has been paid to the consequences of these changes in distribution The 'old' working class culture had its characteristic commercial institutions—the local shop, pub, cinema, betting shop, and so on. These once familiar elements of the 'community' have also changed with the dynamics of post-war capitalism. Like the industrial sector, they too have been subject to concentration, rationalization—and disappearance.

The local geography of cultural institutions, and the networks of socia! relations which they make possible, has changed dramatically. Occasionally, the consequences of this wholesale destruction of stable geographical patterns has generated a wish to revive 'community'. Through the 1960s and '70s, both local and central government confronted urban blight, 'pockets of poverty', vandalism and other symptoms of decay with a deep sense of nostalgia for 'community spirit'.

The transition from the old to the new working class is not an abrupt one, and the trajectory of the inner cities highlights some of the processes of change. Residual groups of the old (and white) working class were left behind in the rush of affluence. Often the elderly, with patterns of life dominated by the lessons and habits of the past, regulated by the routines of shop, pub, club or chapel, preserved--with little choice—some of the old ways, even as the conditions which had supported that culture were eroded around them. As Coates and Silburn (1970) put it in their study of St Anne's in Nottingham, they were the' forgotten Englishmen'—although they were also the' forgotten Englishwomen'. Working class culture always had its defensive features—the mutual supports of'us' againt 'them'-

but under these conditions, these defences became both increasingly desperate and nostalgic. The mobility and hedonism of affluence sat uncomfortably with the locatedness and respectability which were powerful themes ofinter-war working class culture. In, but not of, this world of change, the old was defended because the new had no meaning or reality.

These defensive traditions came to coexist uneasily with more assertive cultural forms which developed in the inner cities. Some areas experienced 'gentrification'—the rehabilitation of old terraces by the new middle classes: mobile, in search of housing but scornful of the new suburbia and its visage of bland respectability. Habitat lived next door to the Coop. This mix of youth and age was one form ofrecomposition of the inner cities but race was its other—more visible—axis. Just as the inter-war working class had found the location and density of the inner city suitable conditions for a defensive and supportive culture, so loo ethnic cultures, confronted by the increasingly overt racism of white British society, found a suitable base there, and began to develop their own cultural institutions. This intersection of'new' ethnic cultures with the old geography of class is not coincidental. In the 1950s, migrant labour was wanted for those 'old' sectors of the British economy (textiles, clothing, heavy manufacturing) where the demand for cheap labour to feed the fires of Britain's industrial expansion was at its highest.

Necessarily, then, their economic character as cheap labour meant their distribution followed the pattern of the old industries and their urban centres: Manchester, Leeds, Bradford, the East and West Midlands, and London. Similarly, within these cities, the residential and social forms for ethnic groups became the old inner-city areas.

Excluded from council housing and from the 'good' areas of private housing through a variety of racist mechanisms, they found that some of the old housing of the inner cities (like jobs in the old industries) was available because it was being left behind by the mobile and affluent sections of the white workingclass.

In the Handsworths, Toxteths and Brixtons, the decline of' traditional' working class culture took place alongside the growth of ethnic cultures. The decaying Methodist chapel was overlooked by the new Sikh temple; local cinemas not rescued by Mecca or Ladbrokes to be bingo halls were reopened to show Asian films; the groups of youths hanging about on street corners became dreadlocked West Indians rather than white skinheads. In these ways, then, the diversity of ethnic cultures came to be overlaid on the geographical pattern of old working class cultures.

Meanwhile, what of the newer, more mobile working class—off to the New Towns and new estates? It would be misleading to see this movement as a simple and total surrender to the new privatized consumer society. No doubt there was an increase of' home-based leisure' and a rise in the purchase of consumer goods. But the conditions of class are not so easily removed. Affluence, while it lasted, was always fragile—hedged by the possibilities of sickness and unemployment, and underpinned as much by the increase in hire purchase as by rising incomes.

The 'never-never' was the route to the consumer paradise. Here, too, traces of the old ways persisted--drink. gambling and sport remained the predominant male working class leisure pursuits. And new forms of cultural difference were constructed out of the new conditions: working men's clubs 'modernized' themselves in the image of market taste in decor and entertainment while stubbonly retaining their distinctive membership structures, internal democracy and institutionalized male dominance.

In surveys of the new leisure society, the working class consistently participate less--retaining a degree of cultural separation even in television viewing. BBC-2 still encounters a stubborn resistance to Reithian cultural improvement, only winning working class audiences for its forays into televised sport. The patterns of life are being remade, but hardly completely anew. Aspects of old structures persist along with traces of old cultures. Women's changing position in paid employment has not undermined their primary responsibility for domestic labour--the re-creation of labour power.

Equally, the increasing mechanization of domestic labour has not abolished its gender specificity. The home remains the site of the woman's ' labour of love'. It is too early to tell what 'new' cultures may emerge around this new geography. The old patterns have been undermined, but new ways of life have not yet solidified into distintive cultural shapes.. This is not particularly surprising. The old cultures stood in the way of a very powerful vision of Britain's future—an expanding economy, full employment, high wages and the consumer paradise. But that dream has turned into a nightmare. The economic, geographical and cultural consequences of boom do not fit easily with those accompanying recession. One example may suffice to show the speed of this turn around.

In the 1960s, the search for the 'new' working class focused on the car workers. They—the affluent, mobile, mass-production workers-were the vanguard of change and progress. Fifteen years after *The Affluent Worker* was published, this identification of car workers as the advance guard seems true only in the most cruelly ironic way. The car industry (and its workers) has faithfully reflected the dynamics of British capitalism: closures, redundancies, increasing proportions of production transferred overseas, the substitution of new technology for labour, and British plants becoming assembly rather than manufacturing enterprises. Halewood, Dagenham, Longbridge, Ryton, Linwood, Luton and the new estates surrounding them may still be the 'vanguard'-a signpost to the future. But the future which they now represent is a long way from the self-confident assertion of affluence and the end of class. It is a future of decline.

Towards Future

This chapter began by tracing the way in which class and class cultures were embedded in the geography of British capitalism. Nation, region and locality had contours along which stable and well understood class boundaries took shape in the inter-war period. I have tried to show how changes in post-war British capitalism have transformed these once stable patterns. The economic composition of class has been changed; the patterns of class cultures have been disrupted; and the geography of class and culture has been restructured. Nor have these processes of economic, cultural and spatial rearrangement come to a standstill. No new certainties have emerged to replace the old familiar patterns. Nostalgia for a sense of place that once was is built into this dynamic of change, as the disruption of the familiar is softened by a sense of the past. The new geography of difference may acquire its own stable cultural patterns boundaries, but as yet, the contemporary cultures of difference are an unstable and uncertain mix of the old and the new.

In the process of change, there is no simple replacement of old cultures by new cultural forms. As was indicated earlier, old patterns are defended and maintained in the face of destabilizing social experience. Resistance to the threat of the new takes many forms, ranging from trade union defence of jobs and working practices in the face of new technology to the maintenance of church and chapel attendance in the face of growing secularization. At other points, old cultural patterns cross-cut and slide into emerging ones--the growing involvement of women trade unionists in previously male-dominated organizations, or the growth of new religious activity (the temples and West Indian fundamentalist Christianity) alongside the established practices of church and chapel. This process is one in which resistance and negotiation are combined in the transformation of cultural forms and practices. In all of these processes, place—the social geography of Britain--has played a central role. Economic change has had a distinctive geographical shape. As before, Britain is divided along the North-South axis but it is not the same pattern of difference. Scotland, Wales, the Northeast, the Northwest, the Midlands are marked by the process of decline, while the new hi-tech industries in the Southeast and East promise a new economic miracle. The old urban centres have borne the brunt of the dynamics of both expansion and contraction—losing jobs and population to the peripheries and leaving the inner city as the desolate symbol of economic and social decline.

4

Economic Geography

Economic geography is the study of the location, distribution and spatial organization of economic activities across the Earth. It focuses on the location of industries and retail and wholesale businesses, on transportation and trade, and on the changing value of real estate. Courses in economic geography may cover such topics as transportation, agriculture, industrial location, world trade, and the spatial organization and function of business activity.

Areas of Study

The distribution of economic activities on Earth is influenced by many environmental, social, political, historical and other factors. Geology can affect resource availability, geomorphology the cost of transportation, and the quality of soil land use decisions. Climate can influence natural resource availability (forestry products) and location or type of agriculture. The social and political institutions that are unique to a particular region also have an impact on economic decisions.

Economic geography research focuses on the study of spatial aspects of economic activities on various scales. The distance to the city (or Central business district) plays a significant role in economic decisions of firms while other factors such as access to the sea and the presence of raw materials like oil affects the economic conditions of countries. Singapore, for example, occupies a key position as a seaport, while the wealth of Saudi Arabia depends almost entirely on oil.

In today's world location and distribution and character of economic activities is much influenced by globalization. States and their borders play less significant role as many countries tend to eliminate the effects of borders and deepen the mutual cooperation of border regions that are often economically marginal and underdeveloped. The best example is the creation of European Union.

Approaches to Study

- *Behavioral economic geography* examines the cognitive processes underlying spatial reasoning, decision making, and behavior of firms and individuals.
- *Historical economic geography* examines history and the development of spatial economic structure.
- *Regional economic geography* examines the economic conditions of particular regions or countries of the world. It deals with economic regionalization as well.
- *Theoretical economic geography* focuses on building theories about spatial arrangement and distribution of economic activities.

Subdivision

Thematically economic geography can be divided into these subdisciplins:

- Geography of agriculture
- Geography of industry

- Geography of services
- Geography of transportation
- and others

However, their areas of study may overlap with another geographical sciences or may be considered on their own.

History of Economic Geography

In the history of economic geography there were many influences coming mainly from economics and geographical sciences.

First traces of the study of distribution of economic activities on Earth can be found in Strabo's *Geographika* written almost 2000 years ago.

During the period known in geography as environmental determinism notable (though later much criticized) influence came from Ellsworth Huntington and his theory of climatic determinism. Valuable contributions came from location theorists such as Johann Heinrich von Thünen or Alfred Weber.

Other influential theories are Walter Christaller's Central place theory, the theory of core and periphery or Pred's matrix.

Contemporary economic geographers tend to specialize in areas such as location theory and spatial analysis (with the help of geographic information systems), market research, geography of transportation, land or real estate price evaluation, regional development and others.

5

Development Geography

Development geography is the study of the Earth's geography and its relationship with economic development. It is very closely related to economic geography and development economics. This branch of human geography is intertwined with resource distribution and consumption, demographics, soil quality, topography, climate and natural disasters. As the world is divided into the More Economically Developed Countries (MEDCs, also called Developed countries and First World Countries) and Less Economically Developed Countries (LEDCs, also Developing countries) wealth (both money and materials resources) is unequally distributed among the world's population with large consequences for people's lifestyles and the environment. In development geography geographers study spatial patterns in development and try to find by what characteristics they can measure economic development. They seek to understand both the geographical causes and consequences of varying development.

Measurement of Development

There are many methods that geographers use to quantify and compare the development of different countries and they have different strengths and weaknesses.

Quantitative Indicators are numerical indications of development.

- GNP per capita measures the value of all the goods and services produced in a country, excluding those produced by foreign companies. This figure is widely used but has many problems. It does not take into account the distribution of the money which can often be extreme unequal as in the UAE where oil money has been collected by a rich elitc and has not flowed to the bulk of the country. GNP does not measure whether the money produced is actually improving people's lives and this is important because in many MEDCs where there are large increases in wealth over time but only small increases in happiness. The figure rarely takes into account the unofficial economy, which includes subsistence agriculture and cash-in-hand or unpaid work, which is often substantial in LEDCs. In LEDCs it is often too expensive to accurately collect this data and some governments intentionally or unintentionally release inaccurate figures. The figure is usually given in US dollars which due to changing currency exchange rates can distort the money's true street value so it is often converted using purchasing power parity (PPP) in which the actual comparative purchasing power of the money in the country is calculated (fOOk).
- Social (demographic) indicators include the birth rate, death rate and fertility rate.
- Health indicators include nutrition (calories per day, calories from protein, percentage with malnutrition), infant mortality and population per doctor.

- Economic indicators include unemployment rates, energy consumption and percentage of GNP in primary industries.

Composite Indicators combine several quantitative indicators into one figure and generally provide a more balanced view of a country. Usually they include one economic, one health and one educational indicator.

- The HDI (Human Development Index) is now the most widely used composite indicator. A number is calculated between 0 and 1 taking into account the most important measures: GNP per capita, the adult literacy rate, the school enrollment rate and life expectancy. It was started by the United Nations in 1990 to replace GNP as a more accurate way of measuring development. A HDI between 1 and 0.8 is considered high, 0.8 and 0.6 is considered medium and 0.6 to 0.4 is considered low.
- Other composite measures include the PQLI (Physical Quality of Life Index) which was a precursor to the HDI which used infant mortality rate instead of GNP per capita and rated countries from 0 to 100. It was calculated by assigning each country a score of 0 to 100 for each indicator compared with other countries in the world. The average of these three numbers makes the PQLI of a country.
- The HPI (Human Poverty Index) is used to calculate the percentage of people in a country who live in relative poverty. In order to better differentiate the number of people in abnormally poor living conditions the HPI-1 is used in developing countries, and the HPI-2 is used in developed countries. The

HPI-1 is calculated based on the percentage of people not expected to survive to 40, the adult illiteracy rate, the percentage of people without access to safe water, health services and the percentage of children under 5 who are underweight. The HPI-2 is calculated based on the percentage of people who do not survive to 60, the adult functional illiteracy rate and the percentage of people living below 50% of median personal disposable income.

- The GDI (Gender-related Development Index) measures gender equality in a country in terms of life expectancy, literacy rates, school attendance and income.

Qualitative Indicators include descriptions of living conditions and people's quality of life. They are useful in analysing features that are not easily converted to numbers such as freedom and security.

DATA EXAMPLE

HDI rank	*Country*	*GDP per capita (PPP US$) (HDI) 2002*	*Human development index (HDI) value 2002*
3	Australia	28,260	0.946
72	Brazil	7,770	0.775
147	Zimbabwe	2,400	0.491

Source: UN Human Development Report (HDR). The UN Human Development Report is the standard for most development statistics worldwide and provides them freely on their website.

Geographic Variations in Development

It is important to grasp just how unequally distributed the world's wealth is. Economic growth over the past 50 years has been impressive. In general richer countries are improving at a faster rate than the poorer countries because they have more money to invest in improvements.

> *"Global wealth also increased in material terms, and during the period 1947 to 2000, average per capita incomes tripled as global GDP increased almost tenfold (from $US3 trillion to $US30 trillion)... Over 25% of the 4.5 billion people in LEDCs still have life expectancies below 40 years. More than 80 countries have a lower annual per capita income in 2000 than they did in 1990. The average income in the world's five richest countries is 74 times the level in the world's poorest five, the widest it has ever been. Nearly 1.3 billion people have no access to clean water. About 840 million people are malnourished."* — Codrington, Stephen. *Planet Geography 3rd Edition* (2005) Page 97.

The most famous pattern in development is the North-South divide. The North-South divide is the divide which separates the rich North or the developed world, from the poor South. This line of division is not as straightforward as it sounds and splits the globe into two main parts.

The "North" in this divide is regarded as being North America, Europe, Russia, Japan, Australia and New Zealand. The countries within this area are generally the more economically developed. The "South" therefore encompasses the remainder of the Southern Hemisphere, mostly consisting of LEDCs. Another possible dividing line is the Tropic of Cancer with the exceptions of Australia and New Zealand. It is critical to understand that the status of countries is far from static and the pattern is likely to become distorted with the fast development of certain southern countries, many of them NICs (Newly Industrialised Countries) including Thailand, Brazil, Malaysia, Mexico and others. These countries are experiencing sustained fast development on the back of growing manufacturing industries and exports.

Most countries are experiencing significant increases

in wealth and standard of living. However there are unfortunate exceptions to this rule. Noticeably the former Soviet Union has experienced major disruption of industry in the transition to a market economy. Many African nations have recently experienced reduced GNPs due to wars and the AIDS epidemic, including Angola, Congo, Sierra Leone and others. Arab oil producers rely very heavily on oil exports to support their GDPs so any reduction in oil's market price (currently unlikely) can lead to rapid decreases in GNP. Countries which rely on only a few exports for much of their income are very vulnerable to changes in the market value of those commodities and are often derogatively called banana republics. Many developing countries do rely on exports of a few primary goods for a large amount of their income (coffee and timber for example), and this can create havoc when the value of these commodities drops, leaving these countries with no way to pay off their debts.

Within countries the pattern is that wealth is more concentrated around urban areas than rural areas. Wealth also tends towards areas with natural resources or in areas that are involved in tertiary (service) industries and trade. This leads to a gathering of wealth around mines and monetary centres such as New York, London and Tokyo.

Causes of Inequality

There are many reasons why some countries develop faster than others. They can be placed under 5 headings using the mnemonic SHEEP, but there is much overlap:

Social

The more money a country has, the more it can spend on health care, education and birth control. Social traditions that discourage birth control increase birth rates and impede the economic development of poor countries. Different societies value hard work, material gain and social cohesion differently and this will clearly have an effect on growth and efficiency. Also, there is the vicious "poverty cycle." This is when the general attitude in developing countries is to have large amounts of children, as they are a guarantee, almost insurance for later life. However, it is hard to cope with large amounts of children, this leads to over population in a country and therefore more poverty. The cycle then starts again.

Historical

Historically colonialism has probably had the largest influence on development. It channeled resource wealth towards Europe and North America at the expense of many African, South American and Asian countries which did not receive reasonable prices for their goods. European colonizers build strong industries from this wealth while not investing in such development in their colonies. At the end of colonialism many countries were left without the social, economic or political structures that encouraged development so poverty became entrenched. In many cases artificial borders were drawn which did not reflect the desires of the local inhabitants, leading to civil wars or social instability. Other historical influences can include incompetent governments or a retention of tribal lifestyles that prevented countries from developing economically.

Economic

Countries with resources such as iron ore, oil and coal are likely to develop industrially more easily because they do not have to import these resources, leading to debt. Their extraction and sale create jobs and transport systems while giving certain countries trade and political leverage over others. The resources can also earn large sums of money in trade, allowing a country to invest in other industries. Many European nations developed during the industrial revolution on the back of coal and iron industries. However, the fact that many resource-rich (oil in particular) African and Middle-Eastern nations have not developed economically while their resources are mined demonstrates that these factors are not sufficient in themselves. Often kleptocracies develop around these industries and grow very rich while investing little in the country's population itself. Nigeria and Saudi Arabia are two examples of this. In fact the wealth generated can often help to entrench an incompetent dictatorship in power or even lead to destructive resource wars as has occurred in Africa.

Environmental

Natural hazards including flooding, droughts, earthquakes, volcanoes, storms, hurricanes, diseases, illnesses and pests all prevent economic development. Large natural disasters can set countries back greatly in their economic development, as in periodic flooding of Bangladesh. Volcanos and floods can often have both positive and negative effects as they bring in nutrient rich sediment. Areas around volcanoes and flooding deltas are often heavily populated, as in Egypt, Bangladesh and Indonesia. Diseases such as malaria which thrive in tropical climates and AIDS which is endemic in Africa prevent people from working and create an economic burden on society. Pests such as locusts reduce agricultural output and make it more difficult for a country to earn sufficient money to escape from subsistence agriculture. Reliable sources of water are necessary for productive agriculture and to a lesser extent industry. Human induced environmental problems include desertification, salinity, water pollution, land clearing and many more. Desertification is caused by poor land management removing the nutrients necessary for plant growth. It is a worldwide problem with massive consequences for the countries it affects. Salinity is caused by poor irrigation techniques. Water pollution from industry can include acids and bases, poisonous minerals and material with a high BOD which cause algal blooms.

This pollution makes it more difficult for a populations to access fresh water. Logging initially brings in investment but often land with trees removed is of far reduced agricultural value and is vulnerable to desertification. Logged rain forests are especially vulnerable to mineral leeching due to high rainfall and often become worthless. As tourism is now a major source of income for most LEDCs it is necessary to care for natural resources which can bring in this long-term source of wealth.

Political

Countries are far more likely to develop when they have stable governments that well macro-manage the economy and invest in national infrastructure, trade, environmental management and avoid civil wars. The rule of law is necessary to make investors feel confident enough to send their money into a country. Clear legal rules on property ownership provides people the opportunity to use the property for collateral on loans for capital development, or sell some property to achieve capital for other endeavors.

Ensuring workers rights can mean that workers receive more money and can pull themselves out of poverty but can have the effect of encouraging multinationals to leave the country and seek other countries with fewer restrictions. It has been suggested that good governance is a prerequisite for economic development and most aid-donor countries now recognise that much of their money has achieved nothing due to corrupt recipients. For this reason standards of governance are now requirements for most aid money. Unfortunately, many governments, especially in Africa, are either unable or unwilling to help their own countries develop and without this support a country can rarely progress.

6

Population Geography

Population Geography is a division of Human Geography. It is the study of the ways in which spatial variations in the distribution, composition, migration, and growth of populations are related to the nature of places. Population geography involves demography in a geographical perspective. It focuses on the characteristics of population distributions that change in a spatial context. Examples can be shown through population density maps. A few types of maps that show the spatial layout of population are chloropleth, isoline, and dot maps. Demography studies:

- Study of people in their spatial distribution and density
- Increase or decrease in population numbers
- The movements and mobility of populations
- Occupational Structure
- Grouping of people in settlements
- The way from the geographical character of places e.g. settlement patterns
- The way in which places in turn react to population phenomena e.g. immigration

All of the above are looked at over space and time.

It must be noted that the boundary between population geography and demography is becoming more and more blurred.

7

Urban Geography

Urban geography is the study of areas which have a high concentration of buildings, infrastructure and economic activity as well as a high population density. There will not be any agricultural activity in an urban environment. It can be considered a part of the larger field of human geography, however, it can often overlap with other fields such as anthropology and physical geography. Urban geographers seek to understand why cities are where they are, what function they serve, the hierarchy that exists between them and why they develop with the form that they have. Urban geography also involves planning city expansion and improvements while taking into account the human and environmental impacts of the change. Urban geography forms the theoretical basis for a number of professions including urban planning, retail store site selection, real estate development, crime pattern analysis, logistical analysis, among others.

There are essentially two approaches to Urban geography. The study of problems relating to the spatial distribution of cities themselves and the complex patterns of movement, flows and linkages that bind them in space. Studies in this category are concerned with the *city system*.

Secondly there is the study of patterns of distribution and interaction within cities, essentially the study of their inner structure. Studies in this category are concerned with the *city as a system*. A succinct way to define urban geography that recognizes the link between these two approaches within the subject is then, that "urban geography is the study of cities as systems within a system of cities."

Site and Situation

Site describes the location of a city with respect to its soil, water supply and relief, while situation describes the surrounds of the city in terms of other settlements, rivers, mountains and communication. Locations for cities are usually chosen for good reasons. Benefits of certain locations can include:

- A wet area: water is a constant necessity for urban areas and is difficult to transport. For this reason many cities are located near or adjacent to rivers.
- A dry area: in wet areas a dry area offers protection from flooding and marshland.
- Easy access to building materials: stone, wood or clay are necessary for the costruction of cities and are difficult to transport long distances.
- A strategic defensive position: historically many cities have been constructed on high ground in order to make attack more difficult and to give a good view of surrounding land. River meanders are also used as partial moats.
- A fuel supply: most cities were initially constructed near wood for burning and cooking. Today many cities are constructed near coal, oil and gas mines to make use of those resources.

- A food supply: cities need some nearby land to be suitable for animal grazing or crop growing
- A travel intersection point and bridging points: it is often useful for a city to be located at the intersection of rivers, roads or train lines in order to facilitate travel and trade. Bridging points are shallow areas that allow easy construction of bridges.
- Shelter and aspect: it is desirable to construct cities located on the side of a slope that is protected from incoming winds, and in a direction that receives maximum sun exposure.

The City System

The large, developed nations all have one defining factor in common. They are all highly urbanized. Before 1850 no country could claim to be predominantly urbanized and at the turn of the century, only Great Britain could be regarded as such. Yet now, over 100 years later many countries are urbanized and less-developed nations are rapidly becoming so.

Urbanization is the process whereby society is transformed from an essentially rural one to a predominatly urban one. Its most visible expression in the landscape is the growth of cities and an increase in their number, size and importance. Also, urbanization process can be closely allied with economic colonization. Urbanization, however is not just the growth of cities but rather it is a complex change in economic, social and political thinking. As cities grow, new definitions have been introduced for census purposes in which the concept of the "extended city" is commonly used. Definitions such as the urban area based on the physical extent of the built-up area and minimum requirements of population size and density as well as the metropolitan area, based on urban population size and commute patterns as a measure of spatial integration give us a much better indication of the population size of cities than the use of political entities like the municipality. One of the definitions that have raised much interest is that based on the concept of urban field which is a new form of urban habitat of relatively low density involving a good transportation system and a broad array of economic, social and recreational oppurtunities. Each urban field is centered on and dominated to a certain extent by a metropolitan area of at least 200,000-300,000 people. It's outer limits can be defined by two criteria: (1) The maximum time or distance that most people are prepared to commute. (2) The time or distance that most people are prepared to spend traveling to or from weekly or weekend recreational activities. Many of these urban fields come into contact with each other and they interact.

Cities do not and cannot exist in isolation. Whether you view the developments in transportation and communication as prerequisites or agents for the growth of cities, the concentration of surplus products in cities necessitated linkages initially with their surrounding regions and later as settlement proceeded westward and a spatial division of labour accompanied the geographic specialization of production, with other regions and cities. In order for the surplus to be concentrated at particular locations linkages were necessary and cities became nodes in networks comprising of the movement of goods, services, materials, people, money, credit, investment and information. These various forms of flows, movements, transactions and linkages are collectively referred to as spatial interaction and acts as the key to the city system. Interaction then plays a number of crucial roles in shaping the form and structure of city systems, as it does in the internal structure of cities. Four roles the city system plays are particularly important. Firstly, in the same way that the market economy is integrated through price-fixing mechanisms, so interaction performs a spatial integrating role. Second, interaction permits differentation of functional specialization of cities within the city system. Third and most important, interaction is the medium of spatial organization. Lastly, interaction is extremelly important in bringing about change and the reorganization of spatial relationships within the city system.

The diffusion of innovations is the ideal example of the function of the urban hierarchy. Large cities are more the most likely place for invention and innovation to occur and spread or diffuse to other cities below it in the hierarchy. As some cities grew and become more important they climb the hierarchy, with major metropolitan centers at the top and towns, villages and hamlets at the bottom.

Cities as Centers of Manufacturing and Services

Cities differ in their economic makeup, their social and demographic characteristics and the roles they play within the city system. These differences can be traced back to regional variations in the local resources on which, growth was based during the early development of the urban pattern and in part the subsequent shifts in the competitive advantage of regions brought about by changing locational forces affecting regional specialization within the framework of the market economy. Recognition of different city types necessitates their classification, and it is to this important aspect of urban geography that we now turn. Emphasis is on *functional town classification* and the basic underlying dimensions of the city system.

The purpose of classifying cities is twofold. On the one hand, it is undertaken in order to search reality for hypotheses. In this context, the recognition of different types of cities on the basis of, for example, their functional specialization may enable the identification of spatial regularities in the distribution and structure of urban functions and the formulation of hypotheses about the resulting patterns. On the other hand, classification is undertaken to structure reality in order to test specific hypotheses that have already been formulated. For example, to test the hypotheses that cities with a diversified economy grow at a faster rate then those with a more specialized economic base, cities must first be classified so that diversified and specialized cities can be differentiated.

The simplest way to classify cities is to identify the distincive role they play in the city system. There are three distinct roles.

1. *Central places* functioning primarily as service centers for local hinterlands.
2. *Transportation* cities performing break-of-bulk and allied functions for larger regions.
3. *Specialized-function* cities are dominated by one activity such as mining, manufacturing or recreation and serving national and international markets.

The composition of a cities labor force has traditionally been regarded as the best indicator of functional specialization, and different city types have been most frequently indentified from the analysis of employment profiles. Specialization in a given activity is said to exist when employment in it exceeds some critical level.

The relationship between the city system and the development of manufacturing has become very apparent. The rapid growth and spread of cities within the heartland-hinterland frameworkafter 1870 was conditioned to a large extent by industrial developments and that the decentralization of population within the urban system in recent years is related in large part to the movement of employment in manufacturing away from the traditional industrial centers. Manufacturing is found in nearly all cities, but its importance is measured by the proportion of total earnings received by the inhabitants of an urban area. When 25 percent or more of the total earnings in an urban region are derived from manufacturing, that urban areas is arbitrarily designated as a manufacturing center.

The location of manufacturing is affected by myriad economic and non-economic factors, such as the nature of the material inputs, the factors of production, the market and transportation costs. Other important influences inclue agglomeration and external economies, public policy and personal preferences. Although it is difficult to evaluate precisely the effect of the market on the location of manufacturing activities, two considerations are involved: the nature of and demand for the product and transportation costs.

8

Behavioral Geography

Behavioral geography is an approach to Human Geography that examines human behavior using a disaggregate approach. Behavioral Geographers focus on the cognitive processes underlying spatial reasoning, decision making, and behavior.

Because of the name it is often incorrectly assumed to have its roots in behaviourism. Due to the emphasis on cognition, this is clearly not the case.

The cognitive processes include environmental perception and cognition, wayfinding, the construction of cognitive maps, place attachment, the development of attitudes about space and place, decisions and behavior based on imperfect knowledge of one's environs, and numerous other topics.

The approach adopted in behavioral geography is closely related to that of psychology, but draws on research findings from a multitude of other disciplines including economics, sociology, anthropology, transportation planning, and many others.

9

Historical Geography

Historical Geography is the study of the: Human; Physical; Fictional; Theoretical and "Real" geographies of the past.

Historical Geography studies a wide variety of issues and topics. A common theme is the study of the geographies of the past and how a place or region changes through time. Many historical geographers study geographical patterns through time, including how people have interacted with their environment, and created the cultural landscape.

Historical geography seeks to determine how cultural features of various societies across the planet emerged and evolved, by understanding their interaction with their local environment and surroundings.

For some in the United States, the term *historical geography* has a more specialized meaning: the name given by Carl Ortwin Sauer of the University of California, Berkeley to his program of reorganizing cultural geography (some say all geography) along regional lines, beginning in the first decades of the 20th century. To Sauer, a landscape and the cultures in it could only be understood if all of its influences through history were taken into account: physical, cultural, economic, political, environmental. Sauer stressed regional specialization as the only means of gaining sufficient expertise on regions of the world. Sauer's philosophy was the principal shaper of American geographic thought in the mid-20th century. Regional specialists remain in academic geography departments to this day. But some geographers feel that it harmed the discipline; that too much effort was spent on data collection and classification, and too little on analysis and explanation. Studies became more and more area-specific as later geographers struggled to find places to make names for themselves. These factors may have led in turn to the 1950s crisis in geography, which raised serious questions about geography as an academic discipline in the United States.

10

International State System and Related Issues

> The spirit of nationalism springs from the deepest of human emotions. It rises from the yearning of men to be free of foreign domination, to govern themselves. It springs from a thousand rills of race, of history, of sacrifice and pride in national achievement. In our own country, does not the word "America" stir something deeper within us than mere geography? Does not the suffering and the sacrifice of our forebears who fought for our independence flash in our minds with every mention of the word "America"?
>
> —*Herbert Hooever*

It was being felt past few years there had been increasing conflicts between the U.S.A. and several of her main political allies over the extraterritorial assertion of U.S. jurisdiction. Complaints, mainly from European countries and in particular Britain, about the application of U.S. 'long-arm' legislation outside the U.S.A. have led to retaliation against the application of U-S. laws and to U.S. justifications and assertions that some European regulation of international business is also extraterritorial in scope. From the point of view of international lawyers, technical questions of public and private international law are involved, concerning the devising of adequate positive rules for the allocation of regulatory jurisdiction. From the political point of view, policy conflicts are involved which raise the question of whether the Western alliance can maintain a common policy towards such questions as the use of trade sanctions against the Soviet bloc in a period of economic depression (Woolcock, 1982).

Both of these approaches assume that the international state system is essentially a functional one, and that provided the correct technical solutions can be found, it is merely a matter of resolving and accommodating national policy differences. A different perspective on the matter is provided by looking at the literature on the internationalisation of capital and the growth of the multinational corporation. Some 10 years ago, Robin Murray pointed to the growing "territorial non-coincidence" between an increasingly interdependent international economic system and the traditional capitalist (or socialist) nation-state.

Murray based himself on the then nascent research on multinational corporations, which proved a major growth industry in the 1970s and now even has its very own United Nations specialized agency, the Commission on Transnational Corporations. Paralleling the famous remark by Kindleberger, "the national state is just about through as an economic unit" (Kindleberger, 1969, p. 207), Murray posed the question "whether... national capitalist states will continue to be the primary structures within the international economic system, or whether the expanded territorial range of capitalist production will require the parallel expansion of co-ordinated state functions" (Murray, 1971, p. 86). He received a rapid, perhaps over-hasty reply from Bill Warren, who in addition to making some good points about the limitations of Murray's theoretical and empirical treatment of the internationalization of capital, was rash enough to prophesy that since the contradictions between capital and the state are essentially non-antagonistic new international regulatory measures would quickly be forthcoming: "the tax authorities are rapidly getting control of the internal transfer price problem and it is clearly not going to be long before the central bankers, international organizations and State policy-making bodies chain down the Euro-dollar monster so that it is no longer available to do the bidding of large firms" (Warren, 1971, p. 88).

Murray's provocative question sparked a considerable amount of subsequent analysis and discussion, both on the nature and the implications of the internationalization of capital, as well as the theory of capital and the state It was readily apparent that Murray's essentially structural-functionalist view of the relationship between capital and the state was responsible for the starkness of the alternative he posed: either capital would outgrow the nation-state and lay the necessary basis for new co-ordinated interstate or supranational structures, or its growth would be contained within the boundaries of existing or merged nation-states. Nevertheless, much of his work on multinationals has been seminal, and his raising of the question of territoriality was important and has not been followed up adequately.

Subsequent debates have emphasized that there is a contradictory process in which the national state is increasingly involved in intervention to ensure the social and economic processes of expanded reproduction of capital, yet at the same time these processes are increasingly transcending the nation-state. The internationalization not only of the circuits of commodity-capital and money-capital but also productive-capital is creating an increasingly internationally integrated world economy and social structure. However, this does not take place as a smooth process of symmetrical interpenetration of capitals to be followed eventually by a merging of social patterns and political superstructures. The intenationalization of production itself entails an internationalized socialization of productive labour as well as international patterns of commodity circulation, both of which are as much social as economic processes, and involve developing patterns of international class formation and conflict. These developments and the different specific forms they take often owe as much to ideological, legal and even military interventions by states and through international state structures, the specific way in which U.S. antitrust law was interpreted and enforced after 1940 played a major part in shaping the characteristic form of U.S. multinational capital using wholly-owned foreign subsidiaries. Thus the state has both been a contributor to the process of internationalization, as well as being affected by it. In Europe, where the international state system originated and which has also been the source of many of the impulses for its subsequent transformations, the development of the European Community involves new and unique processes for interstate integration, and yet the national states still play a vital role, inter-state conflicts continue, and the creation of a single unified super-state is not envisaged by even the most dedicated Europeanist. Yet the impulse to European integration has been as much political as economic.

Jurisdiction Sovereignty

Here an attempt is made to examine, the growing problems of Jurisdictional definition and conflict between the main capitalist states, especially the United States and Europe. It is not surprising that Jurisdictional conflicts have occurred mainly between the most powerful states, and in relation to economic regulation, for it is through such forms of economic regulation that the main capitalist states have been involved in shaping the patterns of internationalization of capital. These assertions of national state power in relation to the structuring of international business, and the conflicts and attempts at co-ordination that have ensued, have been in many ways more important than the other more prominent and formalized mechanisms of international law.

The increasingly dense network of international organizations of diverse kinds and powers as well as the proliferation of treaties and international arrangements and agreements covering thousands of matters major and minor, have grown in the last thirty years both to express and to mask the increasing powerlessness of the dominated peoples of the world.

However, the problem of jurisdiction is more clearly symptomatic of the crisis of international law and of the international system. The principle of territoriality of jurisdiction is the cornerstone of the international system based on the nation-state. The transition from the personal sovereign to an abstract sovereignty of public authorities over a defined territory was a key element in the development of the capitalist international system, since it provided a multifarious framework which permitted and facilitated the global circulation of commodities and capital. The independent and equal sovereign nation-state is therefore a fetishized form of appearance, for the world system is not made up of an aggregation of compartmentalized units, but is rather a single system in which state power is allocated between different territorial entities.

Positivist definitions of state sovereignty and jurisdiction are more or less adequate for liberal forms of regulation, since they can rely on the separation of prescription and enforcement that characterizes such liberal forms. Hence, standard approaches to jurisdiction distinguish between the jurisdiction to prescribe and the jurisdiction to enforce. Jurisdiction to prescribe can be fairly broad in scope and involve a degree of overlap, indeed this may be inevitable. It has been accepted that jurisdiction is not limited to the territory but can also include jurisdiction to regulate the activities of nationals outside the territory.

Even territorial jurisdiction involves overlap, since it requires an assessment of the place where the acts to be regulated 'take place'. This question is traditionally resolved by asking where the 'constituent elements' of a regulated activity take place.

Thus the classic textbook example of a person in state A shooting a person in state B is used to illustrate the point that jurisdiction depends on whether the rule being applied concerns the acts and intention of the person shooting or the nature of the damage caused to the victim. It is therefore enforcement that acts as the limit to Jurisdictional claims, since the actual organization of states as public authorities with defined powers over specific territory means that to be effective a claim to regulate must either involve persons or property within that territory, or must be acceptable to another state which can assist with enforcement.

The territoriality of enforcement is sometimes expressed by saying that administrative or executive jurisdiction is territorial. This approach inevitably comes under great pressure both from the increasing international interdependence especially of economic activity, and from the transcending of liberal forms of state regulation to more direct interventionist, in which there is far less separation between prescription and enforcement of rules.

Increasing international interdependence not only means that more than one state is frequently likely to be involved with a particular activity; it also makes it easier for a state interest in jurisdiction to be made effective through actual control within the territory of some person or property connected with the activity. Thus a multinational company can be fined through a subsidiary or permanent establishment in a country even in respect of actions which it claims took place elsewhere.

Of course, conversely it is possible for a multinational to arrange for particular physical actions, such as the signing of a contract or the meeting of a board of directors, to take place wherever it might be convenient.

It is such considerations that have led to the move to redefine the territoriality of jurisdiction in terms of effects' instead of a physical notion of where actions take place. Such a redefinition is necessary to prevent the sheer evasion of regulations, for example, by arranging for transactions to take place elsewhere. On the other hand, the effects doctrine can in principle involve very broad Jurisdictional claims. [...] Academic commentators have shown an increasing awareness of the inadequacy of the traditional approach to territorial jurisdiction. Akehurst (1972--1973) acknowledges that as long as the 'effects' doctrine is limited to primary effects, it could even be a better means of keeping Jurisdictional claims within reasonable bounds than the 'constituent elements' approach (p. 155).

Lowe, however, goes further, and concludes that the principle of territorial allocation of sovereignty is now unworkable, and calls boldly for a refinement of the concept of sovereignty in international law, so that it can accommodate both notions of the independence of states and of the increasing interdependence of states, without losing its coherence as a legal principle.

The Sherman Act

An ideological rallying-cry much used by U.S. policy-makers in the inter-rational arena is the strong commitment of U.S. law to the regulation of business to ensure competition for the benefit of the consumer and society as a whole. Certainly, modern competition law was born out of the populist movement against the big trusts that quickly came to dominate U.S. business in the last part of the nineteenth-century (Josephson, 1934). While the increasing strength of working class organizations, especially the rapidly growing unions, led to direct and fierce conflicts with big capital, the broader social opposition to big capital among farmers, consumers and smaller property owners of all kinds, led to the passing of the Sherman Act in 1890: The broad scope of the Act, its use against some of the Trusts, and even some initial court decisions seemed to hold out the hope that it could be used to dismantle the Trusts. But this evaporated with its interpretation by the courts as a pro-competition law based on the 'rule of reason', the development of new devices for company incorporation by lawyers, and the emergence of a corporatist consensus between big business and the state embodied in the setting up of the state regulatory commissions.

Thereafter, the Sherman Act was caught in the basic contradiction that bedevils all competition law, since striking down agreements between firms that are held to restrict competition has the effect of laying them open directly to market forces whose tendency is towards the concentration and centralization of capital by the elimination of weaker firms and growth or merger and acquisition.

The Sherman Act was broadly stated to apply to all combinations or contracts in restraint of trade or monopolies affecting trade or commerce within the U.S.A. or *with foreign nations*. In the first case that arose in relation to foreign trade. Justice Holmes gave a landmark judgment whose echoes have been in dissonance with virtually every subsequent opinion by U.S. judges and writers. In *American Banana v. United Fruit Co.* (1909) 213 U.S. 347, the case involved a dispute between two American entrepreneurs with interests in banana plantation operations in Central America, and the plaintiff complained that the defendant had used a variety of predatory tactics to drive him out of the business, including bribery of officials of the government of Costa Rica which had seized his railway (used to export the bananas).

Justice Holmes found the plaintiff's case based on the "startling proposition " that acts causing damage which took place entirely outside the United States could be governed by an Act of Congress, and he stated firmly that the legality of an act must be determined by the law of the place where it took place. Although Justice Holmes' opinion was expressed in sweeping terms, its actual effects were subsequently limited....

In the first fifty years' of its life the application of the Sherman Act to international trade affecting the United States, potentially broad, was in practice confined mainly to monopolization of foreign sources of raw materials in order to push up import prices into the U.S.A. Before 1940 there were only a dozen cases begun by the U.S. authorities relating to international commerce, and all were in this category. Yet the period 1890-1940 was one in which world trade was dominated by cartels, which often had the connivance of governments. The major large firms in the main industrialized countries attempted to control international trade and investment in many industries through various type of cartel agreement.

These typically involved the controlling of levels of production and the allocation of markets between member firms, each keeping control of its home markets and those in dependent economies. It was not until the late 1930s that the U.S. authorities initiated a policy of attacking this type of international cartel. This anticartel policy formed part of the wartime planning for a new framework for post-war international economic relations. Freedom for big firms to compete by selling and investing in each other's markets was seen as the key to avoiding the nationalistic politico-economic rivalries and state protectionisms of th- prewar period.

Thus, the planning of the postwar international institutions which would guarantee a liberalized system of trade, payments and investment (the International Trade Organisation, established in stunted form as the G.A.T.T., the I.M.F, and the World Bank) had to be preceded by the elimination of restraints on international competition embodied in the cartels. After 1940 there began a veritable flood of cases involving some of the main prewar cartels, in which American firms had been involved with European and other companies, especially in chemicals, electrical equipment, and high-technology industries, such as metal alloys and equipment such as gyroscopes and optical instruments. The gathering of information for many of these cases was facilitated by the fact that under wartime regulations the U.S. assets of enemy firms were taken over by a U.S. custodian. Most of these cases resulted in the cartel arrangements being abrogated by agreement, often embodied in a 'consent decree' approved by the Courts.

The most visible legal results of this new policy were the postwar landmark court judgments, of which the *Alcoa* case is the foremost. This case involved the prewar aluminium cartel set up under the umbrella of a Swiss corporation by French, German, Swiss and British monopolies and with the participation of Arthur V. Davis and the Mellons who dominated the U.S. market through Alcoa (the Aluminium Corporation of America) and who had set up a Canadian corporation, largely it seemed to enable them to participate discreetly in the cartel. Since the court held that the existence of a common group of connecting shareholders was not sufficient to ignore the separate legal identity of Alcoa and the Canadian company, it was faced with a conspiracy made up entirely of non-U.S. firms and whose meetings had all taken place outside the U.S.A., even though, of course, it affected the U.S. economy, as part of the world economy. Justice Learned Hand however hesitated little in holding the Sherman Act to apply, on the grounds that "it is settled law that any state may impose liabilities, even upon persi 'ns not within its allegiance, forconduct outside its borders which the state reprehends and these liabilities other states will ordinarily recognise".

This dictum initiated the so-called 'effects' doctrine of jurisdiction. As applied to antitrust law this policy had a significant effect on postwar developments in foreign investment. The broad effect of this policy, which was backed up by strong court judgments, was to invalidate the involvement of any U.S. firm in any arrangement or agreement having theeffect of limiting competition, which included not only the classic international cartel but even a joint venture to set up a company abroad.

This policy of application of the U.S. laws had a

significant effect on the form taken by the postwar expansion of large American firms in world markets. The sweeping away of the cartels removed any legal barrier to U.S. firms wishing to make incursions into foreign markets, and the doubts cast on the legality of any form of joint venture made a significant contribution to the feature that the postwar foreign expansion of U.S. based firms typically took the form of the 100%-owned foreign subsidiary abroad could be said to involve a 'conspiracy' (between the parent company and its legally separate offspring) and could be argued to 'affect U.S. trade' by e.g. reducing exports from the U.S.A. in favour of local production, in practice the U.S. authorities did not take this view and started no action on this basis. It was during the 1950s that foreign firms and governments began to object to the application of U.S. antitrust law in circumstances which it was alleged involved 'extraterritoriality'.

In many cases the invalidation of agreements under U.S. law affected only the private legal rights of the parties, and the requirements of the enforcement of the U.S. assertion of jurisdiction meant that non-U.S. firms could attempt to evade the effects of the legal cases, if they were willing to do no business in the U.S.A. However, the power of U.S. authorities over U.S. firms meant that the continuation of any arrangements with them was impossible. However, no direct legal conflict arose from the U.S. assertion of jurisdiction unless third party rights or foreign state regulation were involved. One such case involved the British chemicals giant, I.C.I., whose involvement in a number of chemicals cartels with firms including the U.S. firm Du Pont, had been attacked in the U.S. courts. One aspect of the cartel with Du Pont involved patent-pooling, and I.C.I, assigned British patents that it had received from Du Pont to an affiliated company which it part-owned, British

Nylon Spinners. The New York court had ordered the mutual return of the patents that had been exchanged by I.C.I. and Du Pont, but the decree (drawn up no doubt by acute negotiations between Wall Street firms) had included a proviso that nothing it required should oblige a defendant to act contrary to any laws or orders of a foreign state "or instrumentality thereof". British Nylon Spinners brought an action against I.C.I, to enforce its rights under the assignment of the British patent, and Lord Denning was only too happy to proclaim that "the writ of the United States does not run in this country", and enforce those assigned rights. He also carefully pointed to the savings clauses in the New York court's decree, which effectively prevented I.C. I. from being subjected to contradictory court orders.

The assertion of jurisdiction to regulate international business by various U.S. agencies led to increasing conflicts, in particular in relation to matters such asshipping liner conferences. These led various governments to pass legislation attempting to prevent U.S. authorities from carrying out investigations which it was claimed infringed the jurisdiction of other states. For example, the British Shipping Contracts and Commercial Documents Act of 1694 empowered the government to prohibit those carrying on business in the U.K. from complying with foreign government measures which appeared to "constitute an infringement of the jurisdiction which, under international law, belongs to the U.K.". Attempts were made through bodies such as the O.E.C.D. to coordinate western countries' policies to matters such as shipping liner conferences.

Simultaneously, the U.S. also attempted to modify the expression of its jurisdictional claims in ways that most clearly recognise the interests of other states in the regulation of international business. To clarify the applicability of the antitrust law to international business and attempt to mollify criticism, the Antitrust Division of the Department of Justice in 1977 produced a detailed *Antitrust Guide for International Operations.* At the same time, two lower court decisions have reviewed in detail the Jurisdictional problem and adopted a modified approach. In particular. Judge Choy in the *Timberlane* case enunciated a "Jurisdictional rule of reason" which required courts carefully to weigh the foreign relations impact of an assertion of jurisdiction over international business. On the other hand, some writers have argued that the introduction of the "Jurisdictional rule of reason" by U.S. courts has failed to appease other states because the balancing of policy interests of national states in regulating a particular activity "will usually reflect an understandable bias in favour of the forum's policy" (Maier, 1982). Maier argues that the weighing of contracts in determining whether the assertion of jurisdiction is "reasonable" should not be in terms of competing local law policies, but should be informed by the common needs of the international system [...], However, this approach does assume that there is some sort of political or ideological basis for the formulation of generally acceptable principles that express the "common needs of the international system".

Practically, the increasingly acute crisis of capital has led to further competitive conflicts between giant firms and protectionist interventions by states. The most notable occurred when the U.S. firm Westinghouse defaulted on fixe.l-price contracts to supply uranium to nuclear power stations it had built, and alleged that world uranium prices had been pushed up by a secret cartel of uranium producers. It instituted private treble-damage actions under the Sherman Act against the mining companies, including the

British-based RTZ. claiming billions of dollars in damages, and the U.S. Attorney-General began a grand jury investigation. Other governments, some of whom such as Canada and Australia were reported to have backed the cartel, moved to block the U.S. proceedings. The British government, largely as a result of the Westinghouse affair as well as the American action against North Atlantic shipping conferences, moved to replace the 1964 Act with the Protection of Trading Interests Act 1980.[...] The basic provisions of the Act are designed to apply not merely to an infringement of U.K. jurisdiction, but to enable retaliation, in concert with other states if necessary, against an extraterritorial assertion of jurisdiction which threatened U.K.. interests. Thus persons (including companies) carrying on business in the U.K. could be given instructions, or could be permitted to recover the penal element of multiple damage judgments, even if the assertion ofjurisdictioin complained of did not infringe that of the U.K. at all, so long as the Minister considered U.K. trading interests were involved.

This involves two separate possibilities. One is that the jurisdiction of a third state is infringed in circumstances which involve damage to British trading interests. The other is that the initial assertion of jurisdiction, although it is 'extraterritorial' does not involve an infringement of a jurisdiction which could be said to be *exclusively* British. In other words, it provides the basis for a retaliation in defence of British trading interests against an assertion by another state of jurisdiction to regulate international activities that might be considered to fall within the *concurrent* jurisdiction of *both* states.

Thus the comments by the U.S. government that the British Act itself involves some 'extraterritoriality' (U.S.

Government, 1979) slightly misses the point. The 1980 British Act is important because it abandons the attempt to define mutually exclusive spheres of jurisdiction between states under international law. Instead it recognizes that where one state unilaterally regulates an activity that cannot be said to take place entirely within its territory, other states must establish a competing or conflicting regulation if they wish to compel an attempt at international coordination.

Conflicts Over Embargo Policies

During the same period, conflicts have occurred over other areas of 'extraterritorial' assertions of U.S. jurisdiction. These have mainly involved the application of U.S. business regulations to foreign companies owned or controlled by U.S. citizens or U.S. companies - that is to say essentially. foreign subsidiaries of U.S. multinationals. One significant area has been U.S. embargo policies, initially against the Soviet bloc and China, then Cuba, and more recently the financial sanctions against Iran.

The Cold War Embargoes

The history of U.S. attempts to orchestrate western economic warfare against communist nation dates back to the period 1947-50 (Adler-Karlsson, 1968). Initially, the emphasis was on achieving the economic isolation of the Soviet bloc through co-ordinated export controls by the U.S.A. and its allies. The Export Control Act 1949 stated it to be U.S. policy to use trade as a political weapon, by empowering the government to prohibit the export of goods or technical data which might make a significant contribution to the military or economic potential.of any nation or combination of nations threatening the security of the United States.... Using a combination of inducements and threats, the U.S. quickly developed a fairly effective economic isolation of the Soviet bloc between 1948 and 1953, which however also had the effect of reinforcing Stalin's aim of consolidating Soviet domination over eastern Europe. The emphasis of U.S. controls in this period was on exports from the U.S.A., although the Export Control Act also covered the use abroad of U.S.-origin technical data. The strength of U.S. industry and the greater attraction of U.S. aid to countries that had emerged weakened from the World War II could be used to ensure international co-ordination. However, there is some evidence that the extent of European co-operation was made to seem greater than in practice it was, due to the problems of definition and enforcement. Adler-Karlsson in his thorough study of this period concludes that it was largely the threat of the loss of the significant quantities of Marshall Aid that ensured compliance by the European countries.

Amongst the main capitalist countries coordination was established in the form of an obscure but apparently powerful mechanism, the C.G.-Co.Com (Consultative Group-Coordinating Committee) set up in 1950. Although the U.S.A. put the initial proposals for this mechanism to the O.E.E.C., it thought that this would be too open and public a body for the purpose, and C.G.-Co.Com. apparently has existed merely as a 'committee' independent of any formal international organization. Almost total secrecy has surrounded it-even its innocuous name was considered 'classified' as late as 1953. It is thought to be based on no written agreement or treaty, but simply a 'gentlemen's agreement' that the lists of embargoed items, agreed on the basis of unanimity, will be enforced by each member state. Members have been the N.A.T.O. countries with the exception of Iceland and the addition of Japan. The main issue throughout has been disagreement over the concept of'strategic' materials. A strong current in U.S. views has been that any items that help to build the Soviet industrial base assist its preparations for war. Hence, an early disagreement with Europe was over British exports of steel. Vociferous anticommunists in Congress specified that "not even a shirt button" should be exported to the enemy. This led to Khrushchev's satirical response that buttons are obviously the most strategic commodity since without them a soldier would have to fight with one hand, the other being required to hold up his trousers (Adler-Karlsson, 1968, ch. 3). It was after 1953 that the embargo lists apparently began to shrink as the European view that they should cover militarily significant items began to predominate. In 1958 there was a further relaxation.

From that point on, there was a significant gap between the U.S.A.'s own total ban, and the internationally agreed embargo, which involved the continual re-interpretation of the ambiguous notion of 'strategic' items. It was at this point that the potentiality for Jurisdictional conflict became a real one. By the late 1950s foreign direct investment by U.S. corporations had begun to build up substantially, especially in Europe. In any case where there was a disagreement between the U.S. and European governments as to whether a particular commodity was 'strategic' and should be embargoed, a dilemma could arise for the U.S.A. and for U.S. multinationals whether a U.S. corporation prohibited from selling an item to a communist country could do so from one of its foreign subsidiaries.

This was prevented, both by the extension of export control procedure to cover re-exports and by transaction regulations passed under the Trading With the Enemy Act, Section 5b. This was a typical wartime emergency legislative enactment first passed in 1917, giving the President wide powers in times of war or declared national emergency to regulate transactions or expropriate property. It was revised and activated in the Cold War era, when President Truman declared a "national emergency" to be in effect on 16 December 1950, after the outbreak of the Korean War; this was apparently considered to be still in effect up to thirty years thereafter. Finally, the embarrassment of this permanent state of declared national emergency, combined with the need to rationalize U.S. powers of economic retaliation following the Arab oil boycott of 1973-4, led to the passing of a slightly more restricted empowering statute to replace the Trading With the Enemy Act, the International Emergency Economic Powers Act 1977.

In the meantime, foreign assets control regulations were applied under the former Act throughout the 1950s and 1960s, first against eastern Europe and the Soviet Union, then the People's Republic of China, and later North Vietnam and Cuba. These regulations applied not only to persons resident or companies doing business in the U.S.A., but also to any enterprises wherever organized or doing business which could be said to be "owned or controlled" by U.S. nationals or companies. Under the wide definition of'control' adopted, even a minority shareholding in a foreign company could be held to constitute' control'. Thus a U.S. company could be warned that a transaction being entered into by a foreign affiliate, ever if it was jointly owned with foreign investors or companies, constituted e', criminal offence by that affiliate under U.S. law. The Department of the Treasury frankly admitted that the enforcement of these regulations inevitably raised foreign policy questions with America's allies.

Coming to another instance, one notable case arose in France in 1964. Fruehauf-France, a two-thirds subsidiary of the U.S. Fruehauf Corporation, contracted with the French company Berliet to supply trailers for trucks, which were to be supplied to China. It seems that initially this destination was not known in the U.S.A., but once it became known, the Fruehauf parent company was ordered not to allow the trailers to be supplied.

It is not clear whether the U.S. multinational covertly sought a way to evade this order, or whether there was a genuine disagreement between the U.S. parent and the French subsidiary's management but, for whatever reason, the French minority directors applied to the French courts for support. In the event, the Paris Court of Appeals took the view that the interests and personality of the French company should be safeguarded even against those of its majority owners, and it appointed an administrator over the Company for three months to execute the contract. One factor revealed by the Fruehauf case and similar incidents was that the subsidiaries of U.S. multinationals abroad were put at a competitive disadvantage in relation to other firms operating in the same country by the gap between the embargo policies of the U.S.A. and other countries.

This became more acute during the late 1960s and 1970s with increasing competitiveness of international markets and increasing pressure, especially in western Europe, for *detente* with the Soviet Union and the further relaxation of trade bans. The American export control legislation was amended several times to relax the procedures, and to make it easier for the substance of the controls to be varied according to the variable winds of *detente* policy. This culminated in the Export Administration Act of 1979, amending and re-enacting the 1949 Act. Crucially, however, the 1979 Act no longer applied merely to the export of goods and technology *from* the U.S.A., but explicitly covered the "export of any goods or technology subject to the jurisdiction of the United States or exported by any person subject to the jurisdiction of the United States " [Section 4(a)(1)]. This provided the powers to apply export control to foreign subsidiaries of U.S. multinationals, on both 'national security' and' foreign policy' grounds, and even where no national emergency had been declared as required by the International Emergency Economic Powers Act. It was clear that, whatever happened with *detente* policy, the U.S.A. had by no means renounced the economic embargo weapon nor its application as widely as U.S. economic power could reach. Indeed, it is clear that, especially after the Arab oil boycott of 1973-4, U.S. policy-makers were preparing defences against such economic threats to the U.S.A. and for the activation of U.S. economic power on a suitable or necessary occasion.

The Iranian Freeze and the Eurodollar

It seems that this occasion was the Iranian crisis of 1979-80. Whatever the background of the deterioration of political relations between the U.S.A. and Iran, it seems that part of the motivation for the scope and type of sanctions pplied was to test the effectiveness of U.S. control of international economic weapons. In terms of U.S. apologetics, the justification is that economic warfare saves lives. The main problematic aspect of the financial sanctions was their application to Eurodollar deposits made by Iran outside the U.S.A Since the loans made to Iran far exceeded Iran's deposits in the U.S.A. alone it was imperative that the sanctions should apply to all Iran's foreign holdings in the international banking system. It does not seem that any attempt was made to co-ordinate such a policy with the western allies, and it seems unlikely that such a co-ordination could have been possible. Instead. the U.S.

freeze regulations were unilaterally made applicable to dollar deposits in U.S. bank branches abroad as well as in the U.S.A. In addition. the U.S. banks sought the means to put pressure on other international banks to go along with the policy, by creating a sort of 'domino effect'. The U.S. regulations were amended so that U.S. bank branches abroad were not only required to freeze Iranian deposits, but also were permitted to' set off' against these frozen deposits any payments due from Iranian entities. The set-off was not allowed within the U.S.A., apparently because a number of smaller U.S. banks were involved in consortia with loans to Iran without having any Iranian deposits.

Once the freeze was applied by the U.S. banks the Iranian authorities found themselves short of funds to service the extensive loans negotiated under the Shah; they naturally brought legal actions in European courts, notably in London and Paris, to secure the release of their funds there. In the meantime, the refusal of the U.S. banks to accept instructions regarding these deposits meant that Iran was exposed to potential default on her Eurodollar loans, and indeed the U.S. banks that had led the consortia for these loans took steps to trigger the default clauses in the agreements. This action put pressure on the European international banks involved in the loans also to use legal rights of set-off against Iranian deposits, since once default is declared all participating banks are under an obligation to make their best efforts to seek satisfaction of the loan,

These unilateral legal and banking moves were of course accompanied by a major ideological campaign, as well as diplomatic initiatives to rally America's allies. Despite obvious hesitations, what was obtained was at least silent compliance. The actual sanctions applied by European states were minimal since they did not apply to the performance of contracts concluded prior to the hostage seizure (and in Britain a parliamentary move further exempted contracts concluded prior to the legislation itself). But most importantly, no retaliatory action was taken in respect of what had amounted to an assertion of U.S. jurisdiction over the entire international financial system. Carswell's (1982) retrospective evaluation admits that it was the unique and short-lived nature of the emergency that enabled the U.S. freeze to be effective. [...]

Summing Up

Obviously, it is not difficult to discern the political and economic reasons for the specific policy conflicts involved in the conflicts of jurisdiction that have occurred, especially between the U.S.A. and Europe. In addition to the main incidents of which we have given an outline, the Jurisdictional issue has arisen also in relation to other matters, such as the application of U.S. securities and stock exchange regulation. Indeed, no sooner had the gas pipeline matter dropped out of the headlines than another conflict replaced it, this time mainly between the U.S. and the U.K., over the liquidation of Laker Airways, and whether the claim by its liquidator that the airline's collapse was due to a conspiracy between other airlines and airframe makers could be litigated in U.S. courts. The regulation of transatlantic air traffic is clearly an area of concurrent jurisdiction, and the British use once again of the Protection of Trading Interests Act was aimed at blocking U.S. intervention and forcing international negotiations. One of the reasons for the frequent allegations ofextraterritoriality against the U.S.A. has been the rapid expansion of its direct foreign investment, mostly in the form of 100% owned subsidiaries, and especially into Europe, coupled with the American predilection for a formalistic regulation of business, compared to the more informal corporatist concertation often practised in Europe. Nevertheless, European authorities have now been obliged to examine more closely their Jurisdictional policies in relation to business regulation, especially where multinational companies are concerned. Americans have not been slow to point to European regulations which could. have an extraterritorial scope. Indeed, in the aftermath of the pipeline dispute, the European Commission has felt it necessary to attempt to co-ordinate internally its policy on jurisdiction, so that its objections to other states' claims do not contradict too starkly with its own assertions of jurisdiction. However, the contradictions and differences of viewpoint within the Community are so great that it is unlikely that any such guidelines on E.E.C, jurisdiction will actually be achieved, let alone be publicly issued as a Statement of Policy. Other ideas such as a European antiboycott statute, also face big political and legal objections. Nevertheless, the Commission has had to take some note of U.S. objections to extra-territoriality in some of its proposals, however spurious some of the -- objections seem.

Indeed, in general, European regulations applying to international business have been framed much more cautiously than U.S. ones. However, U.S. lobbyists such as the American Chamber of Commerce in Brussels, have objected to European Community proposals on Jurisdictional grounds, notably the Vredeling proposal for a directive to require multi-plant and multinational firms to establish machinery for consultation with worker representatives; and also the seventh Company Law directive requiring consolidated accounts for groups of companies. In both of these cases the question has been whether a parent company outside the E.E.C., could be obliged to furnish information relating to its global activities, in view of its operations in the E.E.C, through a subsidiary or branch. In such cases it is not difficult to see how the obligations can be laid on the subsidiary in such a way as to leave the parent little choice. Another area where there are continuing objections to jurisdiction is competition law. Notably, the current proceeding by the Commission against the computer giant IBM has involved both the company and the U.S. government arguing that the Commission's attempt to make IBM change its marketing practices to make them more 'transparent' and open to competition from plug-compatible competitors goes far beyond the jurisdiction of the E.E.C.

Definitely, it is very hard to maintain that the issue in such cases is about the delimitation of an area of exclusive jurisdiction. In practice, the principle of territoriality can be just as broadly applied as the effects doctrine. The underlying issue is that the increased internationalisation and integration of the world economy has created large areas of overlapping jurisdiction, in which conflicts are found to occur unless national regulations are either to be ineffective, or are co-ordinated. Combined with this, the very different nature of the relationship between capital and the state makes liberal forms of accommodation of Jurisdictional overlaps ineffective. In the past private law Jurisdictional overlaps could be accommodated by conflicts of law rules, and public law was penal and confined to the territory.

Now there is a wide range of state regulation of business that is not considered strictly penal (although penal sanctions are applied very often) and that must be applied to multinational business if it is to be effective. The nature of this form of regulation also negates the other method of accommodating jurisdictional conflicts that is most often suggested: to allow a broad overlap of jurisdiction to prescribe while limiting enforcement strictly to territory, since in any case state enforcement agencies have no powers beyond their borders without agreement.

Apparently, sanctions such as the "denial of export privileges" can be very effective against international business even if only applied territorially. It is not so much the territoriality of enforcement that sets a limit to jurisdiction as rather the creation of conflicts with other states' policies or laws. Sometimes there may be conflicts only with private legal rights which may be privately adjusted by the parties or by provisions such as *force majeure* clauses; or the conflict with law or policy may be (reluctantly) tolerated by the other state for a period (as with the Iran freeze). But the move towards the creation of counteracting administrative powers such as the British provisions for the "protection of trading interests" shows that concurrent regulation of international business is increasingly difficult to tolerate or accommodate through liberal forms. Therefore, what is being attempted increasingly is the setting up of direct processes of co-ordination between state agencies, which provide arenas within which negotiations between those agencies and the giant multinationals can be played out. Within such internationalized neo-corporatist state structures it is likely to be the multinational corporations who will dominate, even more so than on the national plane. Nevertheless, there are serious contradictions for capital involved in the disjunction between economic and political processes of internationalisation, and the redefinition of the international state system is by no means an automatic and straightforward process.

11

Time Geography

Time geography (a.k.a. *time-space geography*) traces its roots back to the Swedish geographer Torsten Hägerstrand who stressed the temporal factor in the spatial human activities. *The time-space path*, devised by Hägerstrand, shows the movement of an individual in the spatial-temporal environment with the constraints placed on the individual by these two factors. Three categories of constraints were identified by Hägerstrand: Authority (limits of accessibility to certain places or domains placed on individuals by for example authorities), capability (limitations of movement by individuals. For example, movement is restricted by biological factors, the need for food, drink and sleep.) and coupling (for how long an individual must interact with other individuals in order to finish a task).

12

Regional Geography

Regional geography is a study of regions of all sizes across the Earth. It has a prevailing descriptive character. The main aim is to understand or define the uniqueness or character of a particular region which consists of natural as well as human elements. Attention is paid also to regionalization which covers the proper techniques of space delimitation into regions.

Regional geography is also considered as a certain approach to study in geographical sciences (similar to quantitative geography or bunch of critical geographies). This approach to study was prevailing during the second half of the 19th and the first half of the 20th century also known as a period of prevailing regional geography paradigm when regional geography took the central position in geographical sciences. It was later criticised for its descriptiveness and the lack of theory (regional geography as an empirical approach of geographical sciences). Massive criticism was leveled against this approach in the fifties and during the quantitative revolution. Main critics were Kimble and Schaefer.

Notable regional geographers were Alfred Hettner from Germany with his concept of chorology, Vidal de la Blache from France with the possibilism approach (possibilism as a softer notion of environmental determinism) and United States geographer Richard Hartshorne with his areal differenciation concept.

In today geographical sciences (or geosciences) there are regionally oriented subdisciplines such as regional economic geography or regional geomorphology. Regional geography is thought as study of the major regions of the world, such as Northern and Latin America, Europe, and Asia and their countries.

Some geographers have also attempted to reintroduce a certain amount of regionalism since the 1980s. These involve a complex definition of regions and their interactions with other scales.

REFERENCES

Kimble, G.H.T. (1951): The Inadequacy of the Regional Concept, in Essays in Geography, edts. L.D.Stamp and S.W. Wooldridge, London, pp. 1951-174.

MacLeod, Gordon and Jones, Martin (2001): Renewing The Geography if Regions. Environment and Planning D 16(9), pp. 669-695.

Schaefer, F.K. (1953): Exceptionalism in Geography: A Methodological Examination, A.A.A.G., Vol. 43, pp. 226-245.

13

Tourism Geography

Tourism Geography (or the Geography of Tourism) is the study of travel and tourism as an industry, as a human activity, and especially as a place phenomenon. From a geographical point of view, tourism consists of the places of tourist origin (or tourist generating areas), tourist destinations (or places of tourism supply), and the relationship (connections) between origin and destination places, which includes transportation routes, business relationships, and traveler motivations.

Geography is fundamental to the study of tourism because tourism is geographical in nature. Tourism occurs in places, it involves movement and activities across geographic space (between places), and it is an activity in which both place characteristics and personal self-identities are formed through the relationships that are created among places, landscapes and people. Physical geography provides the essential background against which tourism places are created, and environmental impacts and concerns are major issues that must be considered in managing the development of tourism places.

Human geography provides an understanding of the social and economic relationships that exist in providing tourism and recreation opportunities and activities, as well as the special meaning that these places have to individuals. A destination's sense of place is often a key element behind tourism development -- after all, without the uniqueness and diversity of places, tourism would be mundane and uninteresting.

14

Strategic Geography

Strategic geography is concerned with the control of, or access to, spatial areas that have an impact on the security and prosperity of nations. Spatial areas that concern strategic geography change with human needs and development; recent examples of spatial areas include oil fields which affect the prosperity of a nation or the Gaza strip. This field is a subset of human geography, itself a subset of the more general study of geography. It is also related to geostrategy. For example, Cairo was built on the banks of the Nile river. Not only did this make for fertile land, but easy access to the rest of Ancient Egypt.

15

Feminist Geography

Feminist geography is an approach to study in human geography which applies the theories, methods and critiques of feminism to the study of the human environment and society.

Its principal focus of study is geographic differences in gender relations and gender equality. In addition to its analysis of the real world, it also critiques existing geographical studies, reflecting on the geographical nature of the female experience.

A simple example would be to consider the way that urban planning has a gender dimension with the expectation that men travel to a distant location for employment while women are involved with child care, basic shopping, and domestic functions in a suburban location. Why have cities evolved in this way, why have New towns been planned in this way, and by whom and with what ideology?

Another illustration of this approach to geography is to study gender differences in terms of personal access, mobility and safety, especially in respect to the design and use of urban space and open places such as public parks and footpaths. The experience of walking alone through a city centre late at night or public park even in daytime is likely to be different for males and females and also for people of different ages, cultures and so on.

In this respect feminist geography is part of a broader postmodern approach to the subject which is not primarily concerned with the development of conceptual theory in itself but rather focuses on the real experiences of individuals and groups in their own localities, upon the geographies that they live in within their own communities.

However, feminist geography is not limited to the local scale. One example of a global topic of feminist geography research is the worldwide migration of women from the Third World to the First World to perform domestic labor and sex work.

REFERENCES

Linda Mcdowell. *Gender, Identity and Place: understanding feminist geographies*, Cambridge : Polity Press, 1999—introduction

Companion to Feminist Geography (Blackwell Companions to Geography), ed. by Joni Seager, Lise Nelson, Blackwell Publishers 2004.

16

Political Geography

Political geography is a field of human geography that is concerned with politics. It is closely related to geopolitics, which is seen as the strategic, military and governmental application of political geographies. It is also closely related to International Relations.

Areas of Study

Political geography is interested in the relationship between political power or poltics and geography. Geography often influences political decisions and vice versa political power influences geographical space. It is often associated with the study of the practices of sovereign states, but it considers politics of all scales, from those of the United Nations to those of day-to-day life.

In particular, then, modern political geography often considers:

- How and why states are organized into regional groupings, both formally (e.g. the European Union) and informally (e.g. the Third World)
- The relationship between states and former colonies, and how these are propogated over time through neo-colonialism

- The relationship between a government and its people
- The relationships between states including international trades and treaties
- The functions, demarcations and policings of boundaries
- How imagined geographies have political implications
- The influence of political power on geographical space
- The study of election results (election geography)

History

Pre-World War Two

The term political geography was first used by Friedrich Ratzel in his book 'Politische Geographie', published in German in 1897. Geopolitics was then coined by the Swede Rudolf Kejell.

The discipline gained attention largely through the work of Sir Halford Mackinder in England and his formulation of the Heartland Theory in 1904. This theory involved concepts diametrically opposed to the notion of Alfred Thayer Mahan about the significance of navies (Mahan coined the term *sea power*) in world conflict. The Heartland theory, on the other hand, hypothesized the possibility for a huge empire to be brought into existence which didn't need to use coastal or transoceanic transport to supply its military industrial complex, and that this empire could not be defeated by all the rest of the world coalitioned against it.

The Heartland Theory depicted a world divided into a *Heartland* (Eastern Europe/Western Russia); *World Island*

(Eurasia and Africa); *Peripheral Islands* (British Isles, Japan, Indonesia and Australia) and *New World* (The Americas). Mackinder claimed that whoever controlled the Heartland would have control of the world. He used this warning to politically influence events such as the Treaty of Versailles, where buffer states were created between the USSR and Germany, to prevent either of the them controlling the Heartland.

Ratzel, at the same time, was creating a theory on states baced around the concepts of Lebensraum and Social Darwinism. This stated that states were 'organisms' that needed sufficient room in which to live. Expansionist inclanations - such as the British of French empires were a result of this. Both these two writers created the idea of a political and geographical science, with an objective God's Eye View of the world.

Pre-World War Two political geography was concerned largely with these issues of global power struggles and influencing state policy. The above theories were both taken on board by German geopoliticians (see Geopolitik) such as Karl Haushofer who - they claim inadvertedly - greatly influenced Nazi political theory. The politics that were legitimated by 'scientific' theories such as a 'nautral' requirement for state expansion are those that were ingaged in during World War Two. Though modern geographers have been more sympathetic to the likes of Haushofer, suggesting that he and his colleagues actually did believe that they were conducting neutral scientific study, it is also almost impossible that they could not have forseen how their results would be used; Haushofer, in particular, actually tutored Rudolf Hess and is anacdotely said to have passed a copy of Ratzel's *Politische Geographie* onto Adolf Hitler, whilst he was writing Mein Kampf.

Cold War Geopolitics

After World War II, the demonization of Nazi geopolitics lead to a fall in the popularity and legitimacy of the subject. Geographers of this time became more engaged in economic geography, and regional geography. Geography as a discipline went through what is called the quantitative revolution in the 1960s, moving it from a descriptive subject to a more rigorous, theoretically-grounded discipline. Ironically, this reflected closer the stances of inter-war geopoliticians, who casted themselves as scientific observers and analysers.

During this time, when geography took a positivist stance, political geography remained a fairly narrow subject. It was still interested in the division of the world into different groupings - theories such as core and periphery were dominant, with a Marxist informed critique of capitalism.

Critical Political Geography

Critical political Geography is mainly concerned with the criticism of traditional political geographies. Focusing on the expansion of geographical knowledge, within each critical argument is generally nested an alternative cause for the issue at hand, such as within Andrew Feenberg's *Critical Theory of Technology*.

Notable Political Geographers

- John Agnew
- Halford Mackinder
- Gearóid Ó Tuathail
- Derek Gregory
- Doreen Massey
- Karl Haushofer

- Friedrich Ratzel
- Lynne Staeheli
- Colin Flint
- Joni Seager
- Yves Lacoste

REFERENCES

Buleon, P.: *The state of political geography in France in the 1970s and 1980s.*, Progress in Human Geography. Vol. 16, No. 1. Edward Arnold, Kent, p. 24 - 40, 1992.

Feenberg, Andrew. *"Critical Theory of Technology."* SFU. 22 May 2006.

17

Children's Geographies

Children's Geographies is an area of study in human geography, studying the places and spaces of children's lives. An interest in children's geographies has developed in academic human geography since the beginning of the 1990s, although there were notable studies in the area before that date.

This development emerged from the realisation that previously human geography had largely ignored the everyday lives of children, who (obviously) form a significant section of society, and who have specific needs and capacities, and who may experience the world in very different ways. Thus children's geographies can in part be seen in parallel to an interest in gender in geography and feminist geography in so much as their starting points were the gender blindness of mainstream academic geography.

Children's geographies rests on the idea that children as a social group share certain characteristics which are experientially, politically and ethically significant and which are worthy of study. The pluralistion in the title is intended to imply that children's lives will be markedly different in differing times and places and in differing circumstances such as gender, family, and class.

Children's geographies is sometimes coupled with, and yet distinguished from the geographies of childhood. The former has an interest in the everyday lives of children; the latter has an interest in how (adult) society conceives of the very idea of childhood and how this impinges on children's lives in many ways. This includes imaginations about the nature of children and the related (spatial) implications.

There are a whole range of focii with children's geographies including children and the city, children and the countryside, children and technology, children and nature, children and globalisation, methodologies of researching children's worlds and the ethics of doing so; see the otherness of childhood.

There is now a journal of Children's Geographies which will give readers a good idea of the growing range of issues, theories and methodologies of this developing and vibrant sub-discipline.

18

Human Geography

Human geography is a branch of geography that focuses on the systematic study of patterns and processes that shape human interaction with the environment, with particular reference to the causes and consequences of the spatial distribution of human activity on the Earth's surface.

Scope

It encompasses human, political, cultural, social, and economic aspects of the social sciences. While the major focus of human geography is not the physical landscape of the Earth (see physical geography) it is hardly possible to discuss human geography without referring to the physical landscape on which human activities are being played out, and environmental geography is emerging as an important link between the two. Human geography is methodologically diverse using both qualitative methods and quantitative methods, including case studies, survey research, statistical analysis, and model building among others.

Disciplines of Human Geography	*Related Fields*
Behavioral geography	Psychology
Cultural geography	Anthropology and Sociology
Development geography	Economic development

Economic geography	Economics
Environmental geography	Environmental science
Feminist geography	Feminism
Historical geography	History
Language geography	Linguistics
Marketing geography	Business
Health geography	Health Science and Epidemiology
Military geography	Geostrategy
Political geography	Political science (including Geopolitics)
Population geography	Demography
Regional geography	Regionalisation
Religion geography	Religion
Social geography	Sociology
Strategic geography	Geostrategy
Urban geography	Urban studies and Planning

REFERENCE

Blij, H.J. de; Murphy, Alexander B. (2000). *Human Geography: Culture, Society, and Space,* 7th edition, John Wiley & Sons..

19

Population Geography

Population Geography is a division of Human Geography. It is the study of the ways in which spatial variations in the distribution, composition, migration, and growth of populations are related to the nature of places.

Population geography involves demography in a geographical perspective. It focuses on the characteristics of population distributions that change in a spatial context. Examples can be shown through population density maps.

A few types of maps that show the spatial layout of population are chloropleth, isoline, and dot maps. Demography studies:

- Study of people in their spatial distribution and density
- Increase or decrease in population numbers
- The movements and mobility of populations
- Occupational Structure
- Grouping of people in settlements
- The way from the geographical character of places e.g. settlement patterns
- The way in which places in turn react to population phenomena e.g. immigration

All of the above are looked at over space and time.

N.B. The boundary between population geography and demography is becoming more and more blurred.

20

Coastal Geography

Coastal geography is the study of the dynamic interface between the ocean and the land, incorporating both the physical geography(i.e coastal geomorphology, geology and oceanography) and the human geography(sociology and history) of the coast. It involves an understanding of coastal weathering processes, particularly wave action, sediment movement and weather, and also the ways in which humans interact with the coast.

The waves of different strengths that constantly hit against the shoreline are the primary movers and shapers of the coastline. Despite the simplicity of this process, the differences between waves and the rocks they hit result in hugely varying shapes.

The effect that waves have depends on their strength. Strong, also called destructive waves occur on high energy beaches and are typical of Winter. They reduce the quantity of sediment present on the beach by carrying it out to bars under the sea. Constructive, weak waves are typical of low energy beaches and occur most during summer. They do the opposite to destructive waves and increase the size of the beach by piling sediment up onto the berm.

Probably the most important effect is longshore drift (LSD), the process by which sediment is continuously moved along beaches by wave action. LSD occurs because waves hit the shore at an angle, pick up sediment (sand) on the shore and carry it down the beach at an angle (this is called swash). Due to gravity, the water then falls back perpendicular to the beach, dropping its sediment as it loses energy (this is called backwash). The sediment is then picked up by the next wave and pushed slightly further down the beach, resulting in a continual movement of sediment in one direction. This is the reason why long strips of coast are covered in sediment, not just the areas around river mouths, which are the main sources of beach sediment. LSD is reliant on a constant supply of sediment from rivers and if sediment supply is stopped or sediment falls into a submarine canals at any point along a beach, this can lead to bare beaches further along the shore.

LSD helps create many landforms including barriers, bay beaches and spits. In general LSD action serves to straighten the coast because the creation of barriers cuts off bays from the sea while sediment usually builds up in bays because the waves there are weaker (due to wave refraction), while sediment is carried away from the exposed headlands. The lack of sediment on headlands removes the protection from waves them and makes them more vulnerable to weathering while the gathering of sediment in bays (where longshore drift is unable to remove it) protects the bays from further erosion and makes them pleasant recreational beaches.

Atmospheric Processes

The strength and direction of waves is almost wholly dependent on wind, but the atmosphere plays a larger role than this.

- Wind coming up the beach, picks up sand and moves it up the beach to form sand dunes.
- Rain hits the shore and erodes rocks and carries weathered material to the shoreline to form beaches.
- Warm weather can make for easier for chemical and biological processes to occur. In tropical areas some plants and animals protect stones from weathering, while others actually eat away at the rocks.
- Weather that is sometimes below the freezing point and sometimes above it results in freeze-thaw weathering, while weather more than a few degrees below freezing point creates sea ice that stops most weathering processes.

Biological Processes

In tropical regions in particular, plants and animals not only affect the weathering of rocks but are a source of sediment themselves. The shells and skeletons of many organisms are of calcium carbonate and when this is broken down it forms sediment, limestone and clay.

Chemical Processes

The main chemical process on beaches is salt-crystal growth. Wind carries salt spray onto rocks, where it is absorbed into small pores and cracks within the rocks. There the water evaporates and the salt crystallises, creating pressure and often breaking down the rock. In some beaches calcium carbonate is able to bind together other sediments to form beachrock and in warmer areas dunerock.

Sea Level Changes

Because the sea level on earth regularly rises and falls due to climatic changes. During cold periods more of the

Earth's water is stored as ice in glaciers while during warm periods it is released and sea levels rise to cover more land. Sea levels are currently quite high, while just 18,000 years ago during the Pleistocene ice age they were quite low. Global warming may result in further rises in the future, which presents a risk to coastal cities as most would be flooded by only small rises. As sea levels rise fjords and rias form. Fjords are flooded glacial valleys and rias are flooded river valleys. Fjords typically have steep rocky sides, while rias have dendritic drainage patterns typical of drainage zones. As tectonic plates move about the Earth they can rise and fall due to changing pressures and the presence of glaciers. If a beach is moving upwards relative to other plates this is known as isostatic change and raised beaches can be formed.

Coastal Landforms

Spits

In case, if the coast suddenly changes direction, especially around an estuary, spits are likely to form. LSD pushes sediment along the beach but when it reaches a turn as in the diagram, the LSD does not always easily turn with it, especially near an estuary where the outward flow from a river may push sediment away from the coast. The area may be also be shielded from wave action, preventing much LSD. On the side of the headland receiving weaker waves, shingle and other large sediments will build up under the water where waves are not strong enough to move them along. This provides a good place for smaller sediments to build up to sea level. The sediment, after passing the headland will accumulate on the other side and not continue down the beach, sheltered both by the headland and the shingle.

Slowly over time sediment simply builds on this area,

extending the spit outwards, forming a barrier of sand. Once in a while, the wind direction will change and come from the other direction. During this period the sediment will be pushed along in the other direction. The spit will start to grow backwards, forming a 'hook'. After this time the spit will grow again in the original direction. Eventually the spit will not be able to grow any further because it is no longer sufficiently sheltered from erosion by waves, or because the estuary current prevents sediment resting. Usually in the salty but calm waters behind the spit there will form a salt marshland. Spits often form around the breakwater of artificial harbours requiring dredging.

Occasionally, if there is no estuary then it is possible for the spit to grow across to the other side of the bay and form what is called a bar, or barrier. Barriers come in several varieties, but all form in a manner similar to spits. They usually enclose a bay to form a lagoon. They can join two headlands or join a headland to the mainland. When an island is joined to the mainland with a bar or barrier it is known as a tombolo. This usually occurs due to wave refraction, but can also be caused by isostatic change, a change in the level of the land (e.g. Chesil Beach).

21

Oceanography

Oceanography (from Ocean + Greek *ãñÜöåéí* = write), also called oceanology or marine science is the study of the Earth's oceans and seas. Oceanographers study a wide range of topics such as plate tectonics to ocean currents to marine organisms. These diverse topics reflect multiple disciplines that oceanographers blend to help us understand Earth's interdependencies: biology, chemistry, geology, meteorology and physics.

Branches of Oceanography

The study of oceanography has five main branches:

- *Marine biology* or *biological oceanography,* the study of the plants and animals (biota) of the oceans and their ecological interaction;
- *Chemical oceanography,* the study of the chemistry of the ocean;
- *Marine geology* or *geological oceanography*, the study of the geology of the ocean floor including plate tectonics;
- *Meteorologic oceanography*, studies interactions between our atmosphere and the ocean's hydrosphere; and

- *Physical oceanography,* the study of the ocean's physical attributes (such as its temperature-salinity structure, waves, tides and currents).

These branches reflect the fact that many oceanographers are first trained in the exact sciences and then focus on applying their interdisciplinary knowledge, skills and abilities to oceanography.

History of Oceanography

Early exploration of the oceans was limited to its surfaces and the few creatures that fishermen brought up in nets, but when Bougainville and Cook carried out their explorations in the South Pacific, the seas themselves formed part of the reports.

James Rennell wrote the first scientific textbooks about currents in the Atlantic and Indian oceans during the late 18th and at the beginning of 19th century. Sir James Clark Ross took the first modern sounding in deep sea in 1840, and Charles Darwin published a paper on reefs and the formation of atolls.

The steep slope beyond the continental shelves was not discovered until 1849. Matthew Fontaine Maury's *Physical Geography of the Sea*, 1855 was the first textbook of oceanography. The 1st successful laying of Transatlantic telegraph cable (August 1858) due to the work of Lieutenant M F Maury confirmed the presence of an underwater "telegraphic plateau" mid-ocean ridge.

After the middle of the 19th century, scientific societies were processing a flood of new terrestrial botanical and zoological information. European natural historians began to sense the lack of more than anecdotal knowledge of the oceans.

Oceanography began as a quantifiable science in 1872, when the Scots Charles Wyville Thompson and John Murray launched the Challenger expedition (1872-76). Other European and American nations also sent out scientific expeditions (as did private individuals and institutions).

Oceanographic institutes dedicated to the study of oceanography were founded. In the United States, these included the Scripps Institution of Oceanography, Woods Hole Oceanographic Institution, Lamont-Doherty Earth Observatory at Columbia University, and the School of Oceanography at University of Washington. In Britain, there is a major research institution: National Oceanography Centre, Southampton.

The first international organization of oceanography was created in 1901 as the International Council for the Exploration of the Sea.

In 1921 Monaco formed the International Hydrographic Bureau (IHB). Then in 1966, the U.S. Congress created a *National Council for Marine Resources and Engineering Development*. NOAA was in charge of exploring and studying all aspects of Oceanography. It also enabled the National Science Foundation to award Sea Grant College funding to multi-disciplinary researchers in the field of oceanography.

Ocean and Atmosphere Connections

The study of the oceans is intimately linked to understanding global warming and other global environmental concerns.

> Our planet is invested with two great oceans;
>
> one visible, the other invisible; one underfoot, the other overhead; one entirely envelopes it, the other covers about two thirds of its surface.

Matthew F. Maury (1855) *The Physical Geography of the Seas and Its Meteorology.*

Notable Oceanographers

- Robert Ballard (ONR Science Explorer)
- VADM W.W. Behrens, Jr. (NOAA)
- Henry Bryant Bigelow
- RADM Richard E. Byrd
- Jacques Cousteau
- Robert S. Dietz
- V. Walfrid Ekman
- Evan Forde
- Gotthilf Hempel
- RADM Grace Murray Hopper USNR
- Charles David Keeling
- Uwe Kils
- Sir James Lovelock (NASA Biosphere)
- Matthew Fontaine Maury (US Naval Observatory)
- Walter Munk
- Tad Murty
- Stefan Rahmstorf
- Johan Sandström
- Yuly Shokalsky
- RADM Charles Dwight Sigsbee (Hydographer)
- Henry Stommel
- Harald Sverdrup
- Mary Sears
- Warren White
- Charles Wilkes (United States Exploring Expedition)

Related Disciplines

- Biogeochemistry
- Biogeography
- Hydrography
- Hydrology
- Limnology
- Meteorology

22

Pedology (Soil Study)

Pedology (from the Greek *pedon*, soil, earth), is the study of soils in its natural environment.[1] It is one of two main branches of soil science, the other being edaphology. Pedology deals with pedogenesis, soil morphology, soil classification.

Introduction

Soil is not only a support for vegetation, but it is also the zone (the pedosphere) of numerous interactions between climate (water, air, temperature), soil life (micro-organisms, plants, animals) and its residues, the mineral material of the original and added rock, and its position in the landscape. During its formation and genesis, the soil profile slowly deepens and develops characteristic layers, called 'horizons', while a steady state balance is approached.

Soil users (such as agronomists) showed initially little concern in the dynamics of soil. They saw it as medium whose chemical, physical and biological properties were useful for the services of agronomic productivity. On the other hand, pedologists and geologists did not initially focus on the agronomic applications of the soil characteristics (edaphic properties) but upon its relation to the nature and history of landscapes. Today, there's an integration of the two disciplinary approaches as part of landscape and environmental sciences.

Pedologists are now also interested in the practical applications of a good understanding of pedogenesis processes (the evolution and functioning of soils), like interpreting its environmental history and predicting consequences of changes in land use, while agronomists understand that the cultivated soil is a complex medium, often resulting from several thousands of years of evolution. They understand that the current balance is fragile and that only a thorough knowledge of its history makes it possible to ensure its sustainable use.

Main Concepts

- Complexity in soil genesis is more common than simplicity.
- Soils lie at the interface of Earth's atmosphere, biosphere, hydrosphere and lithosphere. Therefore, a thorough understanding of soils requires some knowledge of meteorology, climatology, ecology, biology, hydrology, geomorphology, geology and many other earth sciences and natural sciences.
- Contemporary soils carry imprints of pedogenic processes that were active in the past, although in many cases these imprints are difficult to observe or quantify. Thus, knowledge of paleoecology, paleogeography, glacial geology and paleoclimatology is important for the recognition and understanding of soil genesis and constitute a basis for predicting the future soil changes.
- Five major, external factors of soil formation (climate, organisms, relief, parent material and time), and several smaller, less identifiable

ones, drive pedogenic processes and create soil patterns.

- Characteristics of soils and soil landscapes, e.g., the number, sizes, shapes and arrangements of soil bodies, each of which is characterized on the basis of soil horizons, degree of internal homogeneity, slope, aspect, landscape position, age and other properties and relationships, can be observed and measured.
- Distinctive bioclimatic regimes or combinations of pedogenic processes produce distinctive soils. Thus, distinctive, observable morphological features, e.g., illuvial clay accumulation in B horizons, are produced by certain combinations of pedogenic processes operative over varying periods of time.
- Pedogenic (soil-forming) processes act to both create and destroy order (anisotropy) within soils; these processes can proceed simultaneously. The resulting soil profile reflects the balance of these processes, present and past.
- The geological Principle of Uniformitarianism applies to soils, i.e., pedogenic processes active in soils today have been operating for long periods of time, back to the time of appearance of organisms on the land surface. These processes do, however, have varying degrees of expression and intensity over space and time.
- A succession of different soils may have developed, eroded and/or regressed at any particular site, as soil genetic factors and site factors, e.g., vegetation, sedimentation, geomorphology, change.
- There are very few old soils (in a geological sense)

because they can be destroyed or buried by geological events, or modified by shifts in climate by virtue of their vulnerable position at the skin of the earth. Little of the soil continuum dates back beyond the Tertiary period and most soils and land surfaces are no older than the Pleistocene Epoch.

- Knowledge and understanding of the genesis of a soil is important in its classification and mapping.
- Soil classification systems cannot be based entirely on perceptions of genesis, however, because genetic processes are seldom observed and because pedogenic processes change over time.
- Knowledge of soil genesis is imperative and basic to soil use and management. Human influence on, or adjustment to, the factors and processes of soil formation can be best controlled and planned using knowledge about soil genesis.
- Soils are natural clay factories (clay includes both clay mineral structures and particles less than 2 μm in diameter). Shales worldwide are, to a considerable extent, simply soil clays that have been formed in the pedosphere and eroded and deposited in the ocean basins, to become lithified at a later date.

Famous Pedologists

- Olivier de Serres
- Bernard Palissy
- Vasily V. Dokuchaev
- Eugene W. Hilgard
- Hans Jenny

REFERENCES

Buol, Stanley W., F,D. Hole and R.W. McCracken. 1997. Soil Genesis and Classification, 4th ed. Iowa State Univ. Press, Ames ISBN 0-81382-873-2.

Hole Francis D., J.B. Campbell. 1985. Soil landscape analysis. Totowa Rowman & Allanheld, 214 p. ISBN 0-86598-140-X.

Ronald Amundsen. Soil Preservation and the Future of Pedology. Retrieved on 2006-06-08. PDF.

23

Cultural Region

Cultural region is a term used mainly in the study of geography. Distinct cultures often do not limit their geographic coverage inside the borders of a nation state, or to smaller subdivisions of a state. To 'map' a culture, we often have to identify an actual 'cultural region', and when we do this we find that it bears little relationship to the legal borders drawn up by custom, treaties, charters or wars.

There are different kinds of cultural regions that can be delineated. A map of culture that maps 'religion & folklore' may have slightly different shape to one which, in the same region, maps 'dress and architecture'.

A formal cultural region is one in which there is clearly a homogeneous cultural identity on a wide variety of levels. These are often found to map onto specific bioregions, reflecting ancient migrations.

A functional cultural region is one that has been largely brought into being as a function of having a central cultural hub (e.g. a monastery or a TV station).

A vernacular cultural region is that 'mental image' of a region which exists in the minds of outsiders.

An aspirational cultural region is that 'mental image' of a region which exists primarily in the minds of those who would work to create it in the future. For instance, in the nationalist struggles on the Balkan Peninsula under the Ottoman Empire, or in the Jewish diaspora before the creation of Israel, or the Catalan region of Spain.

Such regions may expand or contract over time, and cultural values may also shift.

24

Geopolitics

Geopolitics analyses geography, history and social science with reference to international politics. In other words, it studies the political and strategic significance of geography; in this context, geography is defined in terms of the location, size, and resources of places.

The term was coined by Rudolf Kjellén, a Swedish political scientist, at the beginning of the 20th century. Kjellén was inspired by the German geographer Friedrich Ratzel, who published his book "Politische Geographie" (political geography) in 1897. The term was popularized in English by US diplomat Robert Strausz-Hupé, a faculty member of the University of Pennsylvania.

Halford Mackinder

The doctrine of Geopolitics gained attention largely through the work of Sir Halford Mackinder in England and his formulation of the Heartland Theory in 1904. The doctrine involved concepts diametrically opposed to the notion of Alfred Thayer Mahan about the significance of navies (he coined the term *sea power*) in world conflict. The Heartland theory hypothesized the possibility for a huge empire to be brought into existence in the Heartland, which would not need to use coastal or transoceanic transport to supply its military industrial complex, and that this empire could not be defeated by all the rest of the world coalitioned against it.

The basic notions of Mackinder's doctrine involve considering the geography of the Earth as being divided into two sections, the 'World Island' which comprised Eurasia and Africa, and the 'Periphery', which included the Americas, British Isles, and Oceania. Not only was the Periphery noticeably smaller than the World Island, it necessarily required much sea transport to function at the technological level of the World Island, which contained sufficient natural resources for a developed economy. Also, the industrial centers of the Periphery were necessarily located in widely-separated locations. The World Island could send its navy to destroy each one of them in turn. It could locate its own industries in a region further inland than the Periphery could, so they would have a longer struggle reaching them, and would be facing a well-stocked industrial bastion. This region Mackinder termed the *Heartland*. It was essentially comprised of Ukraine, Western Russia, and Mitteleuropa. The Heartland contained the grain reserves of Ukraine, and many other natural resources. Mackinder's notion of geopolitics can be summed up in his saying "Who rules East Europe commands the Heartland. Who rules the Heartland commands the World-Island. Who rules the World-Island commands the world." His doctrine was influential during the World Wars and the Cold War, for Germany and later Russia each made territorial strides toward the Heartland.

Other Theories

After World War I, Kjellen's thoughts and the term were picked up and extended by a number of scientists: in

Germany by Karl Haushofer, Erich Obst, Hermann Lautensach and Otto Maull; in England, Mackinder and Fairgrieve; in France Vidal de la Blache and Vallaux. In 1923 Karl Haushofer founded the "Zeitschrift für Geopolitik" (magazine for geopolitics), which developed as a propaganda organ for Nazi-Germany.

Anton Zischka published *Afrika, Europas Gemischftaufgabe Tummer* (Africa, Complement of Europe) in 1952, where he proposed a kind of North-South Empire, from Stockholm to Johannesburg.

Since then, the word *geopolitics* has been applied to other theories, most notably the notion of the Clash of Civilizations by Samuel Huntington. In a peaceable world, neither sea lanes nor surface transport are threatened; hence all countries are effectively close enough from one another physically. It is in the realm of the political ideas, workings, and cultures that there are differences, and the term has shifted more towards this arena, especially in its popular usage. Traditionally, it strictly applies to geography's effect on politics.

25

Demography

Demography is the scientific study of human population dynamics. It encompasses the study of the size, structure and distribution of populations, and how populations change over time due to births, deaths, migration and ageing. Demographic analysis can relate to whole societies or to groups defined by criteria such as education, nationality, religion and ethnicity. Most countries' demography is regarded as a branch of either economics or sociology. Formal demography limits its object of study to the measurement of populations processes, while the more broad field of population studies also analyze the relationships between economic, social, cultural and biological processes influencing the population.

The term demographics is often used erroneously for demography, but refers rather to selected population characteristics as used in marketing or opinion research, or the demographic profiles used in such research.

Data and Methods

Demography may rely on the use of large amounts of data, including census returns and vital statistics registers, or incorporate survey data using indirect estimation techniques. The earliest modern census was carried out in the United States in 1790, although the Scandinavian countries had earlier censuses.

In many countries, particularly in the third world, reliable demographic data are still difficult to obtain; census is often equated in the minds of the people with taxation, so the people scatter when a census taker comes around. During the 1980s, for example, the population of Nigeria was widely estimated to be around 101 million people, before it was established to be as little as 89 million people (without adjustment for undercounting) in a census carried out in 1991.

Important Concepts

Important concepts in demography include:

- The crude birth rate, the annual number of live births per thousand people.
- The general fertility rate, the annual number of live births per 1000 women of childbearing age (often taken to be from 15 to 49 years old, but sometimes from 15 to 44).
- Age-specific fertility rates, the annual number of live births per 1000 women in particular age groups (usually age 15-19, 20-24 etc.)
- The crude death rate, the annual number of deaths per 1000 people.
- The infant mortality rate, the annual number of deaths of children less than 1 year old per thousand live births.
- The expectation of life (or life expectancy), the number of years which an individual at a given age can expect to live at present mortality levels.

- The total fertility rate, the number of live births per woman completing her reproductive life, if her childbearing at each age reflected current age-specific fertility rates.
- The gross reproduction rate, the number of daughters who would be born to a woman completing her reproductive life at current age-specific fertility rates.
- The net reproduction ratio is the expected number of daughters, per newborn prospective mother, who may or may not survive to and through the ages of childbearing.

Note that the crude death rate as defined above and applied to a whole population can give a misleading impression. For example, the number of deaths per 1000 people can be higher for developed nations than in less-developed countries, despite standards of health being better in developed countries. This is because developed countries have relatively more older people, who are more likely to die in a given year, so that the overall mortality rate can be higher even if the mortality rate at any given age is lower. A more complete picture of mortality is given by a life table which summarises mortality separately at each age. A life table is necessary to give a good estimate of life expectancy.

Basic Demographic Equation

Suppose that a country (or other entity) contains $Population_t$ persons at time t. What is the size of the population at time $t + 1$?

$$Population_{t+1} = Population_t + Births_t \text{ “ } Deaths_t + Immigration_t \text{ “ } Emigration_t$$

Net migration from time t to $t + 1$:

$$Netmigration_{t+1} = Immigration_{t} \text{ `` } Emigration_{t}$$

Natural increase from time t to $t + 1$

$$Naturalincrease_{t+1} = Births_{t} \text{ `` } Deaths_{t}$$

History

The *Natural and Political Observations ... upon the Bills of Mortality* (1662) of John Graunt contains a primitive form of life table. Mathematicians, such as Edmond Halley, developed the life table as the basis for life insurance mathematics. Richard Price was credited with the first textbook on life contingencies published in 1771. (ref: "Our Yesterdays: the History of the Actuarial Profession in North America, 1809-1979," by E.J. (Jack) Moorhead, FSA, (1/23/10 - 2/21/04), published by the Society of Actuaries as part of the profession's centennial celebration in 1989; followed later by Augustus de Morgan, 'On the Application of Probabilities to Life Contingencies', (1838). (ref: The History of Insurance, Vol 3, Edited by David Jenkins and Takau Yoneyama (1 85196 527 0): 8 Volume Set: (2000) Availability: Japan: Kinokuniya)

At the end of the 18th century, Thomas Malthus concluded that, if unchecked, populations would be subject to exponential growth. He feared that population growth would tend to outstrip growth in food production, leading to ever increasing famine and poverty (see Malthusian catastrophe); he is seen as the intellectual father of ideas of overpopulation and the limits to growth. Later more sophisticated and realistic models were presented by e.g. Benjamin Gompertz and Verhulst.

The Demographic Transition

Contrary to Malthus' predictions (though in line with

his thoughts on moral restraint), natural population growth in most developed countries has diminished to close to zero, without being held in check by famine or lack of resources, as people in developed nations have shown a tendency to have fewer children. The fall in population growth has occurred despite large rises in life expectancy in these countries. This pattern of population growth, with slow (or no) growth in preindustrial societies, followed by fast growth as the society develops and industrialises, followed by slow growth again as it becomes more affluent, is known as the demographic transition.

Similar trends are now becoming visible in ever more developing countries, so that far from spiralling out of control, world population growth is expected to slow markedly in the next century, coming to an eventual standstill. The change is likely to be accompanied by major shifts in the proportion of world population in particular regions. The United Nations Population Division expects the absolute number of infants and toddlers in the world to begin to fall by 2015, and the number of children under 15 by 2025. Demographers at the International Institute for Applied Systems Analysis in Austria expect world population to peak at 9000 million by 2070. Throughout the 21st century, the average age of the population is likely to continue to rise.

26

Geomorphology

Geomorphology is the study of landforms, including their origin and evolution, and the processes that shape them. The underlying question is: Why do landscapes look the way they do? The term is derived from the Greek *ãç, ge*, meaning *Earth*, and *ìïñöÞ, morfé*, meaning form. Geomorphologists seek to understand landform history and dynamics, and predict future changes through a combination of field observation, physical experiment, and numerical modeling. The discipline is practiced within geology, geodesy, geography, archaeology, and civil and environmental engineering. Early studies in geomorphology are the foundation for pedology, one of two main branches of soil science.

Landforms evolve in response to a combination of natural and anthropogenic processes. The landscape is built up through tectonic uplift and volcanism. Denudation occurs by erosion and mass wasting, which produces sediment that is transported and deposited elsewhere within the landscape or off the coast. Landscapes are also lowered by subsidence, either due to tectonics or physical changes in underlying sedimentary deposits. These processes are each influenced differently by climate, ecology, and human activity.

Particular applications of geomorphology include landslide prediction and mitigation, river control and restoration, coastal protection, and assessing the presence of water on Mars.

History

Geomorphology was not originally differentiated from the rest of geography. The first geomorphic model was the *geographical cycle* or the *cycle of erosion*, developed by William Morris Davis between 1884 and 1899. The cycle was inspired by theories of evolution which were first formulated by James Hutton (1726-1797). Concerning valley forms, the cycle was depicted as a sequence by which a river would cut a valley more and more deeply, but then erosion of side valleys would eventually flatten out the terrain again, now at a lower elevation. The cycle could be started over by uplift of the terrain. The model is today considered too much of a simplification to be especially useful in practice.

Walther Penck developed an alternative model in the 1920s, based on ratios of uplift and erosion, but it was also too weak to explain a variety of landforms. G. K. Gilbert was an important early American geomorphologist.

Processes

Modern geomorphology focuses on the quantitative analysis of interconnected processes, such as the contribution of solar energy, the rates of steps of the hydrologic cycle, and plate movement rates from geophysics to compute the age and expected fate of landforms. The use of more precise measurement technique has also enabled processes like erosion to be observed directly, rather than merely surmised from other evidence. Computer simulation is also valuable for testing that a particular model yields results with properties similar to real terrain.

Primary surface processes responsible for most topographic features include wind, waves, weathering, mass wasting, ground water, surface water, glaciers, tectonism, and volcanism.

Fluvial Geomorphology

Rivers and streams are not only conduits of water, but also of sediment. The water, as it flows over the channel bed, is able to mobilise sediment and transport it downstream, either as bedload, suspended load or dissolved load. The rate of sediment transport depends on the availability of sediment itself and on the river's discharge.

As rivers flow across the landscape, they generally increase in size, merging with other rivers. The network of rivers thus formed is a drainage system and is often dendritic, but may adopt other patterns depending on the regional topography and underlying geology.

Glacial Geomorphology

Glaciers, while geographically restricted, are effective agents of landscape change. The gradual movement of ice down a valley causes abrasion and plucking of the underlying rock. Abrasion produces fine sediment, termed glacial flour. The debris transported by the glacier, when the glacier recedes, is termed a moraine. Glacial erosion is responsible for U-shaped valleys, as opposed to the V-shaped valleys of fluvial origin.

Weathering

This results from chemical dissolution of rock and from the mechanical wearing of rock by plant roots, ice expansion, and the abrasive action of sediment. Weathering provides the source of the sediment transported by fluvial, glacial, aeolian, or biotic processes.

Taxonomy

Different geomorphological processes dominate at different spatial and temporal scales. To help categorize landscape scales some geomorphologists use the following taxonomy:

- 1st - Continent, ocean basin, climatic zone (~10,000,000 km^2)
- 2nd - Shield, e.g. Baltic shield, or mountain range (~1,000,000 km^2)
- 3rd - Isolated sea, Sahel (~100,000 km^2)
- 4th - Massif, e.g. Massif Central or Group of related landforms, e.g., Weald (~10,000 km^2)
- 5th - River valley, Cotswolds (~1,000 km^2)
- 6th - Individual mountain or volcano, small valleys (~100 km^2)
- 7th - Hillslopes, stream channels, estuary (~10 km^2)
- 8th - gully, barchannel (~1 km^2)
- 9th - Meter-sized features

Its use, however, is rare and may be misleading—the nature of landscape change may be better viewed as a continuum of coupled processes.

- Base level
- Biogeology
- Bioerosion
- Biorhexistasy
- Coastal erosion
- Drainage system

- Erosion prediction
- Fluvial landforms of streams
- Geologic modeling
- Lithosphere
- Regolith
- Soil
- Soil conservation
- Soil mechanics
- Soils retrogression and degradation
- Stream capture
- Watershed
- Important publications in geomorphology

REFERENCES

Adrian E. Scheidegger, *Morphotectonics*, Springer-Verlag Berlin 2004.

Bernhard Edmaier, *Earthsong*. A collection of breathtaking arial photographs. Phaidon, 2004.

M.J. Selby, *Earth's Changing Surface*, Oxford University Press, 1985

Richard Chorley, Stanley Schumm, and David Sugden, *Geomorphology*. Edition Methuen, 1984.

27

Hydrology

Hydrology (from Greek: Yäñoëoãéá, Yäùñ+Ëoãos, *Hydrologia*, the "study of water") is the study of the movement, distribution, and quality of water throughout the Earth, and thus addresses both the hydrologic cycle and water resources. A practitioner of hydrology is a hydrologist, working within the fields of either earth or environmental science, or civil and environmental engineering.

Domains of hydrology include hydrometeorology, surface hydrology, hydrogeology, drainage basin management and water quality, where water plays the central role. Oceanography and meteorology are not included because water is only one of many important aspects.

Hydrological research is useful in that it allows us to better understand the world in which we live, and also provides insight for environmental engineering, policy and planning.

Branches of Hydrology

Chemical hydrology is the study of the chemical characteristics of water.

Ecohydrology is the study of interactions between organisms and the hydrologic cycle.

Hydrogeology is the study of the presence and movement of water in aquifers.

Hydroinformatics is the adaptation of information technology to hydrology and water resources applications.

Hydrometeorology is the study of the transfer of water and energy between land and water body surfaces and the lower atmosphere.

Hydromorphology

Isotope hydrology is the study of the isotopic signatures of water.

Surface hydrology is the study of hydrologic processes that operate at or near the Earth's surface.

Related Fields

- Aquatic chemistry
- Civil engineering
- Climatology
- Environmental engineering
- Geomorphology
- Hydraulic engineering
- Limnology
- Oceanography

Hydrologic Measurements

The movement of water through the Earth can be measured in a number of ways. This information is important for both assessing water resources and understanding the processes involved in the hydrologic cycle. Following is a list of devices used by hydrologists and what they are used to measure.

- Disdrometer - precipitation characteristics
- Evaporation -Symon's evaporation pan
- Infiltrometer - infiltration
- Piezometer - groundwater pressure and, by inference, groundwater depth (see: aquifer test)
- Radar - cloud properties
- Rain gauge - rain and snowfall
- Satellite
- Sling psychrometer - humidity
- Stream gauge - stream flow (see: discharge (hydrology))
- Tensiometer - soil moisture
- Time domain reflectometer - soil moisture

Hydrologic Prediction

Observations of hydrologic processes are used to make predictions of future water movement and quantity.

Statistical Hydrology

By analysing the statistical properties of hydrologic records, such as rainfall or river flow, hydrologists can estimate future hydrologic phenomena. This, however, assumes the characteristics of the processes remain unchanged.

These estimates are important for engineers and economists so that proper risk analysis can be performed to influence investment dicisions in future infrastructure and to determine the yield reliability characteristics of water supply systems. Statistical information is utilised to formulate operating rules for large dams forming part of systems which include agricultural, industrial and residential demands.

Hydrologic Modeling

Hydrologic models are simplified, conceptual representations of a part of the hydrologic cycle. They are primarily used for hydrologic prediction and for understanding hydrologic processes. Two major types of hydrologic models can be distinguished:

- Models based on data. These models are black box systems, using mathematical and statistical concepts to link a certain input (for instance rainfall) to the model output (for instance runoff). Commonly used techniques are regression, transfer functions, neural networks and system identification. These models are known as stochastic hydrology models.
- Models based on process descriptions. These models try to represent the physical processes observed in the real world. Typically, such models contain representations of surface runoff, subsurface flow, evapotranspiration, and channel flow, but they can be far more complicated. These models are known as deterministic hydrology models. Deterministic hydrology models can be subdivided into single-event models and continuous simulation models.

Hydrologic Transport

Water movement is a significant means by which other material, such as soil or pollutants, are transported from place to place. Initial input to receiving waters may arise from a point source dicharge or a line source or area source, such as surface runoff. Since the 1960s rather complex mathematical models have been developed, facilitated by the availability of high speed computers. The most common pollutant classes analyzed are nutrients, pesticides, total dissolved solids and sediment.

Applications of Hydrology

- Mitigating and predicting flood, landslide and drought risk;
- Designing irrigation schemes and managing agricultural productivity;
- Providing drinking water;
- Designing dams for water supply or hydroelectric power generation;
- Designing bridges;
- Designing sewers and urban drainage system;
- Analyzing the impacts of antecedent moisture on sanitary sewer systems.
- Predicting geomorphological changes, such as erosion or sedimentation.
- Assessing the impacts of natural and anthropogenic environmental change on water resources.
- Assessing contaminant transport risk and establishing environmental policy guidelines.

28

Glaciology

Glaciology is the study of glaciers, or more generally the study of ice and natural phenomena that involve ice. The word *glacier* is derived from the Latin *glacies*, meaning ice or frost.

Glaciology is an interdisciplinary earth science that integrates geophysics, geology, physical geography, geomorphology, climatology, meteorology, hydrology, biology, and ecology. The impact of glaciers on humans adds the fields of Human geography and anthropology. The presence of ice on Mars and Europa brings in an extraterrestrial component to the field.

Overview

Areas of study within glaciology include glacial history and the reconstruction of past glaciation patterns, effects of glaciers on climate and vice versa, the dynamics of ice movement, the contributions of glaciers to erosion and geomorphology, lifeforms that live in the ice, and so forth.

Not surprisingly, glaciology is one of the key areas supported by polar research.

Types

There are two general categories of glaciation which glaciologists distinguish: *alpine glaciation*, accumulations or "rivers of ice" confined to valleys; and *continental glaciation*, unrestricted accumulations which once covered much of the northern continents.

- Alpine - ice flows down the valley of mountainous areas and forms an apron of moving ice onto the plains below. Alpine glaciers tend to make the topography more rugged.
- Continental - an ice sheet found only in high latitudes (Greenland/Antarctica), thousands of square kilometers wide and thousands of meters thick. These tend to smooth out the landscape.

Zones of glaciers

- Accumulation
- Wastage

Movement

Ablation: wastage through evaporation and melting

Arête: an acute ridge where two cirques abut.

Bergshrund: crevasse formed at the head of a glacier, where it breaks away from the mountain face.

Cirque, corrie or cwm: bowl shaped depression excavated by the source of a glacier.

Creep: adjustment to stress at a molecular level.

Flow: movement (of ice) in a constant direction.

Fracture: brittle failure (breaking of ice) under the stress raised when movement is too rapid to be accommodated by creep. It happens for example, as the central part of a glacier movinges faster than the edges.

Horn: spire of rock formed by the headward erosion of a ring of cirques around a single mountain. It is an extreme case of an arête.

Plucking / Quarrying: where the adhesion of the ice to the rock is stronger than the cohesion of the rock, part of the rock leaves with the flowing ice.

Tarn: a lake formed in the bottom of a cirque

Tunnel valley: The tunnel is that formed by hydraulic erosion of ice and rock below an ice sheet margin. The tunnel valley is what remains of it in the underlying rock when the ice sheet has melted.

Glacial Deposits

Stratified

Outwash sand / gravel: from front of glaciers, found on a plain

Kettles: block of stagnant ice leaves a depression or pit

Eskers: steep sided ridges of gravel/sand, possibly caused by streams running under stagnant ice

Kames: stratified drift builds up low steep hills

Varves: thin sedimentary beds (coarse to fine), summer deposits more material and in the winter, less.

Unstratified

Till-unsorted: (glacial flour to boulders) deposited by receding/advancing glaciers, forming moraines, and drumlins

Moraines: (Terminal) material deposited at the end; (Ground) material deposited as glacier melts; (lateral) material deposited along the sides.

Drumlins: smooth elongated hills composed of till.

Ribbed Morainess: large subglacial elongated hills transverse to former ice flow.

REFERENCES

Benn, Douglas I. and David J. A. Evans. *Glaciers and Glaciation*. London; Arnold, 1998.

Hambrey, Michael and Jürg Alean. *Glaciers* 2nd ed. Cambridge University Press, 2004.

Knight, Peter G. *Glaciers* Cheltenham; Nelson Thornes, 1999.

29

Biogeography

Biogeography is the science which deals with patterns of species distribution and the processes that result in such patterns.

The patterns of species distribution at this level can usually be explained through a combination of historical factors such as speciation, extinction, continental drift, glaciation (and associated variations in sea level, river routes, and so on), and river capture, in combination with the area and isolation of landmasses (geographic constraints) and available energy supplies.

History

The theory of biogeography grows out of the work of Alfred Russel Wallace and other early evolutionary scientists. Wallace studied the distribution of flora and fauna in the Malay Archipelago in the 19th Century. With the exception of Wallace and a few others, prior to the publication of *The Theory of Island Biogeography* by Robert MacArthur and E.O. Wilson in 1967 (which expanded their 1963 paper on the same topic) the field of biogeography was seen as a primarily historical one, and as such the field was seen as a purely descriptive one. MacArthur and

Wilson changed this perception, and showed that the species richness of an area could be predicted in terms of such factors as habitat area, immigration rate and extinction rate. This gave rise to an interest in island biogeography. The application of island biogeography theory to habitat fragments spurred the development of the fields of conservation biology and landscape ecology (at least among British and American academics; landscape ecology has a distinct genesis among European academics).

Classic biogeography has been expanded by the development of molecular systematics, creating a new discipline known as phylogeography. This development allowed scientists to test theories about the origin and dispersal of populations, such as island endemics). For example, while classic biogeographers were able to speculate about the origins of species in the Hawaiian Islands, phylogeography allows them to test theories of relatedness between these populations and putative source populations in Asia and North America.

Classification

Biogeography is a synthetic science, related to geography, biology, soil science, geology, climatology, ecology and evolution.

Some fundamentals in biogeography are

- evolution (change in genetic composition of a population)
- extinction (disappearance of a species)
- dispersal (movement of populations away from their point of origin, related to migration)
- range and distribution
- endemic areas.

30

Limnology

Limnology is a discipline that concerns the study of inland waters (both saline and fresh), specifically lakes, ponds and rivers (both natural and manmade), including their biological, physical, chemical, and hydrological aspects. The term *limnology* stems from Greek *limne* (lake) and *logos* (study).

François-Alphonse Forel (1841-1912) established the field with his studies of Lake Geneva. Limnology traditionally is closely related to hydrobiology, which is concerned with the application of the principles and methods of physics, chemistry, geology, and geography to ecological problems.

Organizations

- American Society of Limnology and Oceanography
- Australian Society for Limnology
- European Society of Limnology and Oceanography
- German Society of Limnology (www.dgl-ev.de)
- Italian Association for Oceanology and Limnology (AIOL) (www.aiol.info - *R.Bertoni*)
- The Japanese Society of Limnology

- Societas Internationalis Limnologiae (limnology.org)
- Brazilian Society of Limnology
- New Zealand freshwater Sciences society
- Southern African Society of Aquatic Scientists

Journals

- *Advances in Limnology*
- *Chinese Journal of Oceanology and Limnology*
- *Journal of Limnology* (Journal webpage - *R.Bertoni*)
- *Limnologica*
- Limnology and Oceanography
- Journal of Coastal Research

Resources

- The History of Limnology at the University of Wisconsin-Madison: A digital resource documenting three generations of limnological research in Wisconsin. The collection mainly focuses on the important pioneers of limnology, Dr. Edward A. Birge, Chancey Juday and Arthur D. Hasler, research laboratories, and field equipment. Presented by the University of Wisconsin Digital Collections Center.
- Breaking new waters: a Century of Limnology at the University of Wisconsin.

REFERENCES

A.J.Horne & C.R. Goldman: *Limnology* (1994).

B. Moss, *Ecology of Fresh Waters* (Blackwell, 1998).

Gerald A. Cole, *Textbook of Limnology*, 4th ed. (Waveland Press, 1994)

G.E. Hutchinson, *A Treatise on Limnology*, 3 vols. (1957-1975) - classic but dated.

H.B.N. Hynes, *The Ecology of Running Waters* (1970).

Jacob Kalff, *Limnology* (Prentice Hall, 2001).

Robert G. Wetzel, *Limnology: Lake and River Ecosystems*, 3rd ed. (Academic Press, 2001).

Robert G. Wetzel & Gene E. Likens, *Limnological Analyses*, 3rd ed. (Springer-Verlag, 2000).

31

Geodesy

Geodesy, also called geodetics, is the scientific discipline that deals with the measurement and representation of the earth, its gravitational field and geodynamic phenomena (polar motion, earth tides, and crustal motion) in three-dimensional, time-varying space.

Definition

Geodesy is primarily concerned with positioning and the gravity field and geometrical aspects of their temporal variations, although it can also include the study of the Earth's magnetic field. Especially in the German speaking world, geodesy is divided in geomensuration ("Erdmessung" or "höhere Geodäsie"), which is concerned with measuring the earth on a global scale, and surveying ("Ingenieurgeodäsie"), which is concerned with measuring parts of the surface.

The shape of the earth is to a large extent the result of its rotation, which causes its equatorial bulge, and the competition of geologic processes such as the collision of plates and of vulcanism, resisted by the earth's gravity field. This applies to the solid surface (orogeny; few mountains are higher than 10 km, few deep sea trenches deeper than that.) Quite simply, a mountain as tall as, for example, 15 km, would develop so much pressure at its base, due to gravity, that the rock there would become plastic, and the mountain would slump back to a height of roughly 10 km in a geologically insignificant time. (On Mars, whose surface gravity is much less, the largest volcano, Olympus Mons, is 27 km high at its peak, a height that could not be maintained on Earth.) Gravity similarly affects the liquid surface (dynamic sea surface topography) and the earth's atmosphere. For this reason, the study of the Earth's gravity field is seen as a part of geodesy, called physical geodesy.

Geoid and Reference Ellipsoid

The geoid is essentially the figure of the Earth abstracted from its topographic features. It is an idealized equilibrium surface of sea water, the mean sea level surface in the absence of currents, air pressure variations etc. and continued under the continental masses. The geoid, unlike the ellipsoid, is irregular and too complicated to serve as the computational surface on which to solve geometrical problems like point positioning. The geometrical separation between it and the reference ellipsoid is called the geoidal undulation. It varies globally between $\pm$ 110 m.

A reference ellipsoid, customarily chosen to be the same size (volume) as the geoid, is described by its semi-major axis (equatorial radius) a and flattening f. The quantity $f = (a - b) / a$, where b is the semi-minor axis (polar radius), is a purely geometrical one. The mechanical ellipticity of the earth (dynamical flattening, symbol J_2) is determined to high precision by observation of satellite orbit perturbations. Its relationship with the geometric flattening is indirect. The relationship depends on the internal density distribution, or, in simplest terms, the degree of central concentration of mass.

The 1980 Geodetic Reference System (GRS80) posited a 6,378,137 m semi-major axis and a 1: 298.257 flattening. This system was adopted at the XVII General Assembly of the International Union of Geodesy and Geophysics (IUGG). It is essentially the basis for geodetic positioning by the Global Positioning System and is thus also in extremely widespread use outside the geodetic community.

The numerous other systems which have been used by diverse countries for their maps and charts are gradually dropping out of use as more and more countries move to global, geocentric reference systems using the GRS80 reference ellipsoid.

Coordinate Systems in Space

The locations of points in three-dimensional space are most conveniently described by three cartesian or rectangular coordinates, X,Y and Z. Since the advent of satellite positioning, such coordinate sytems are typically *geocentric*: the Z axis is aligned with the Earth's (conventional or instantaneous) rotation axis.

Before the satellite geodesy era, the coordinate systems associated with geodetic datums attempted to be geocentric, but their origins differed from the geocentre by hundreds of metres, due to regional deviations in the direction of the plumbline (vertical). These regional geodetic datums, such as ED50 (European Datum 1950) or NAD83 (North American Datum 1983) have ellipsoids associated with them that are regional 'best fits' to the geoids within their areas of validity, minimising the deflections of the vertical over these areas.

It is only because GPS satellites orbit about the geocentre, that this point becomes naturally the origin of a coordinate system defined by satellite geodetic means, as the satellite positions in space are themselves computed in such a system.

Geocentric coordinate systems used in geodesy can be divided naturally into two classes:

1. Inertial reference systems, where the coordinate axes retain their orientation relative to the fixed stars, or equivalently, to the rotation axes of ideal gyroscopes; the *X* axis points to the vernal equinox
2. Co-rotating, also ECEF ("Earth Centred, Earth Fixed"), where the axes are attached to the solid body of the Earth. The *X* axis lies within the Greenwich observatory's meridian plane.

The coordinate transformation between these two systems is described to good approximation by (apparent) sidereal time, which takes into account variations in the Earth's axial rotation (length-of-day variations). A more accurate description also takes polar motion into account, a phenomenon currently closely monitored by geodesists.

Coordinate Systems in the Plane

In surveying and mapping, important fields of application of geodesy, two general types of coordinate systems are used in the plane:

1. Plano-polar, in which points in a plane are defined by a distance *s* from a specified point along a ray having a specified direction á with respect to a base line or axis;
2. Rectangular, points are defined by distances from two perpendicular axes called *x* and *y*. It is geodetic

practice -- contrary to the mathematical convention -- to let the x axis point to the North and the y axis to the East.

Rectangular coordinates in the plane can be used intuitively with respect to one's current location, in which case the x axis will point to the local North. More formally, such coordinates can be obtained from three-dimensional coordinates using the artifice of a map projection. It is *not* possible to map the curved surface of the Earth onto a flat map surface without deformation. The compromise most often chosen -- called a conformal projection -- preserves angles and length ratios, so that small circles are mapped as small circles and small squares as squares.

An example of such a projection is UTM (Universal Transverse Mercator). Within the map plane, we have rectangular coordinates x and y. In this case the North direction used for reference is the *map* North, not the *local* North. The difference between the two is called *meridian convergence*.

It is easy enough to "translate" between polar and rectangular coordinates in the plane: let, as above, direction and distance be á and s respectively, then we have

$$x = s \cos \alpha,$$

$$y = s \sin \alpha.$$

The reverse translation is slightly more tricky.

Heights

In geodesy, point or terrain *heights* are "above sea level", an irregular, physically defined surface. Therefore a height should ideally *not* be referred to as a coordinate. It is more like a physical quantity, and though it can be tempting to treat height as the vertical coordinate z, in addition to the horizontal coordinates x and y, and though this actually is a good approximation of physical reality in small areas, it becomes quickly invalid in larger areas.

Heights come in the following variants:

1. Orthometric heights
2. Normal heights
3. Geopotential numbers

Each have their advantages and disadvantages. Both orthometric and normal heights are heights in metres above sea level, whereas geopotential numbers are measures of potential energy (unit: m^2s^{-2}) and not metric. Orthometric and normal heights differ in the precise way in which mean sea level is conceptually continued under the continental masses. The reference surface for orthometric heights is the geoid, an equipotential surface approximating mean sea level.

None of these heights are in any way related to geodetic or ellipsoidial heights, which express the height of a point above the reference ellipsoid. Satellite positioning receivers typically provide ellipsoidal heights, unless they are fitted with special conversion software based on a model of the geoid.

Geodetic Datums

Because geodetic point coordinates (and heights) are always obtained in a system that has been constructed itself using real observations, we have to introduce the concept of a *geodetic datum*: a physical realization of a coordinate system used for describing point locations. The realization is the result of *choosing* conventional coordinate values for one or more *datum points*.

In the case of height datums, it suffices to choose *one* datum point: the reference bench mark, typically a tide gauge at the shore. Thus we have vertical datums like the NAP (Normaal Amsterdams Peil), the North American Vertical Datum 1988 (NAVD88), the Kronstadt datum, the Trieste datum, etc.

In case of plane or spatial coordinates, we typically need several datum points. A regional, ellipsoidal datum like ED50 can be fixed by prescribing the undulation of the geoid and the deflection of the vertical in *one* datum point, in this case the Helmert Tower in Potsdam. However, an overdetermined ensemble of datum points can also be used.

Changing the coordinates of a point set referring to one datum, to make them refer to another datum, is called a *datum transformation*. In the case of vertical datums, this consists of simply adding a constant shift to all height values. In the case of plane or spatial coordinates, datum transformation takes the form of a similarity or *Helmert transformation*, consisting of a rotation and scaling operation in addition to a simple translation. In the plane, a Helmert transformation has four parameters, in space, seven.

A Note on Terminology

In the abstract, a coordinate system as used in mathematics and geodesy is, e.g., in ISO terminology, referred to as a *coordinate system*. International geodetic organizations like the IERS (International Earth Rotation and Reference Systems Service) speak of a *reference system*.

When these coordinates are realized by choosing datum points and fixing a geodetic datum, ISO uses the terminology *coordinate reference system*, while IERS speaks of a *reference frame*. A datum transformation again is referred to by ISO as a *coordinate transformation.* (ISO 19111: Spatial referencing by coordinates).

Point Positioning

Point positioning is the determination of the coordinates of a point on land, at sea, or in space with respect to a coordinate system. Point position is solved by computation from measurements linking the known positions of terrestrial or extraterrestrial points with the unknown terrestrial position. This may involve transformations between or among astronomical and terrestrial coordinate systems.

The known points used for point positioning can be triangulation points of a higher order network, or GPS satellites.

Traditionally, a hierarchy of networks has been built to allow point positioning within a country. Highest in the hierarchy were triangulation networks. These were densified into networks of traverses (polygons), into which local mapping surveying measurements, usually with measuring tape, corner prism and the familiar red and white poles, are tied.

Nowadays all but special measurements (e.g., underground or high precision engineering measurements) are performed with GPS. The higher order networks are measured with static GPS, using differential measurement to determine vectors between terrestrial points. These vectors are then adjusted in traditional network fashion. A global polyhedron of permanently operating GPS stations under the auspices of the IERS is used to define a single global, geocentric reference frame which serves as the "zeroth order" global reference to which national measurements are attached.

For surveying mappings, frequently Real Time Kinematic GPS is employed, tying in the unknown points with known terrestrial points close by in real time.

One purpose of point positioning is the provision of known points for mapping measurements, also known as (horizontal and vertical) control. In every country, thousands of such known points exist in the terrain and are documented by the national mapping agencies. Constructors and surveyors involved in real estate will use these to tie their local measurements to.

Geodetic Problems

In geometric geodesy we formulate two standard problems: the geodetic principal problem and the geodetic inverse problem.

Geodetic principal problem (also: first geodetic problem)

> Given a point (in terms of its coordinates) and the direction (azimuth) and distance from that point to a second point, determine (the coordinates of) that second point.

Geodetic inverse problem (also: second geodetic problem)

> Given two points, determine the azimuth and length of the line (straight line, arc or geodesic) that connects them.

In the case of plane geometry (valid for small areas on the Earth's surface) the solutions to both problems reduce to simple trigonometry. On the sphere, the solution is significantly more complex, e.g., in the inverse problem the azimuths will differ between the two end points of the connecting great circle, arc, i.e. the geodesic.

On the ellipsoid of revolution, solutions in closed form do not exist, so rapidly converging series expansions have traditionally been used, such as Vincenty's formulae.

In the general case, the solution is called the geodesic for the surface considered. It may be nonexistent or non-unique. The differential equations for the geodesic can be solved numerically, e.g., in MATLAB.

Geodetic Observational Concepts

Here we define some basic observational concepts, like angles and coordinates, defined in geodesy (and astronomy as well), mostly from the viewpoint of the local observer.

- The *plumbline* or *vertical* is the direction of local gravity, or the line that results by following it. It is slightly curved.
- The *zenith* is the point on the celestial sphere where the direction of the gravity vector in a point, extended upwards, intersects it. More correct is to call it a <direction> rather than a point.
- The *nadir* is the opposite point (or rather, direction), where the direction of gravity extended downward intersects the (invisible) celestial sphere.
- The celestial *horizon* is a plane perpendicular to a point's gravity vector.
- *Azimuth* is the direction angle within the plane of the horizon, typically counted clockwise from the North (in geodesy) or South (in astronomy and France).
- *Elevation* is the angular height of an object above the horizon, Alternatively zenith distance, being equal to 90 degrees minus elevation.
- *Local topocentric coordinates* are azimuth (direction angle within the plane of the horizon) and elevation angle (or zenith angle) and distance.
- The North *celestial pole* is the extension of the

Earth's (precessing and nutating) instantaneous spin axis extended Northward to intersect the celestial sphere. (Similarly for the South celestial pole.)

- The *celestial equator* is the intersection of the (instantaneous) Earth equatorial plane with the celestial sphere.
- A *meridian plane* is any plane perpendicular to the celestial equator and containing the celestial poles.
- The *local meridian* is the plane containing the direction to the zenith and the direction to the celestial pole.

Geodetic Observing Instruments

The level is used for determining height differences and height reference systems, commonly referred to mean sea level. The traditional spirit level produces these practically most useful heights above sea level directly; the more economical use of GPS instruments for height determination requires precise knowledge of the figure of the geoid, as GPS only gives heights above the GRS80 reference ellipsoid. As geoid knowledge accumulates, one may expect use of GPS heighting to spread.

The theodolite is used to measure horizontal and vertical angles to target points. These angles are referred to the local vertical. The tacheometer additionally determines, electronically or electro-optically, the distance to target, and is highly automated in its operations. The method of free station position is widely used.

For local detail surveys, tacheometers are commonly employed although the old-fashioned rectangular technique using angle prism and steel tape is still an inexpensive alternative. More and more, also real time kinematic (RTK) GPS techniques are used. Data collected is tagged and recorded digitally for entry into a Geographic Information System (GIS) data base.

Geodetic GPS receivers produce directly three-dimensional coordinates in a geocentric coordinate frame. Such a frame is, e.g., WGS84, or the frames that are regularly produced and published by the International Earth Rotation and Reference Systems Service (IERS).

GPS receivers have almost completely replaced terrestrial instruments for large-scale base network surveys. For planet-wide geodetic surveys, previously impossible, we can still mention satellite laser and Very Long Baseline Interferometer (VLBI) techniques. All these techniques also serve to monitor Earth rotation irregularities as well as plate tectonic motions.

Gravity is measured using gravimeters. There are two basic kinds of gravimeters. *Absolute* gravimeters, which nowadays can also be used in the field, are based directly on measuring the acceleration of free fall (for example, of a reflecting prism in a vacuum tube). They are used for establishing the vertical geospatial control. Most common *relative* gravimeters are spring based. They are used in gravity surveys over large areas for establishing the figure of the geoid over these areas. Most accurate relative gravimeters are *superconducting* gravimeters, and these are sensitive to one thousandth of one billionth of the Earth surface gravity. Twenty-some superconducting gravimeters are used worldwide for studying Earth tides, rotation, interior, and ocean and atmospheric loading, as well as for verifying the Newtonian constant of gravitation.

Units and Measures on the Ellipsoid

Geographical latitude and longitude are stated in the units degree, minute of arc, and second of arc. They are *angles*, not metric measures, and describe the *direction* of the local normal to the reference ellipsoid of revolution. This is *approximately* the same as the direction of the plumbline, i.e., local gravity, which is also the normal to the geoid surface. For this reason, astronomical position determination, measuring the direction of the plumbline by astronomical means, works fairly well provided an ellipsoidal model of the figure of the Earth is used.

A geographic mile, defined as one minute of arc on the equator, equals 1,855.32571922 m. A nautical mile is one minute of astronomical latitude. The radius of curvature of the ellipsoid varies with latitude, being the longest at the pole and shortest at the equator as is the nautical mile.

A metre was originally defined as the 40 millionth part of the length of a meridian. This means that a kilometre is equal to (1/40,000) * 360 * 60 meridional minutes of arc, which equals 0.54 nautical miles. Similarly a nautical mile is on average 1/0.54 = 1.85185... km.

Temporal Change

In geodesy, temporal change can be studied by a variety of techniques. Points on the Earth's surface change their location due to a variety of mechanisms:

- Continental plate motion, plate tectonics
- Episodic motion of tectonic origin, esp. close to fault lines
- Periodic effects due to Earth tides

- Postglacial land uplift due to isostatic adjustment
- Various anthropogenic movements due to, e.g., petroleum or water extraction or reservoir construction.

The science of studying deformations and motions in the Earth's crust and the solid Earth as a whole is called geodynamics. Often, also study of the Earth's irregular rotation is included in its definition.

Techniques for studying geodynamic phenomena on the global scale include: satellite positioning by GPS and similar techniques, very long baseline interferometry (VLBI), and satellite and lunar laser ranging. Regionally and locally, precise levelling, precise tacheometers and monitoring gravity change, as well as synthetic aperture radar interferometry (inSAR) from orbit are often used techniques.

International Organizations

- International Association of Geodesy (IAG)
- International Union of Geodesy and Geophysics (IUGG)
- Fédération Internationale des Géomètres (FIG)

University Institutes

The notable university institutes engaged in geodesy include:

- The *Institut für Erdmessung* in Hannover, Germany - which specialises in astro-geodetic zenith cameras and geoid computations for many European countries
- Division of Geodesy - Royal Institute of Technology (KTH) Stockholm, Sweden

- The *Institut für Theoretische Geodäsie* in Bonn (Geodesy, Radio astronomy and GPS)
- The *Institut für Astronomische und Physikalische Geodäsie* in Munich, southern Germany.
- The Austrian *Institute for Geodesy and Geophysics* at the TU Vienna (astro-geological geoid, IGS and VLBI)
- The Swiss *Geodetic Institute* at the ETH Zürich (geophysical geodesy, GPS etc.)
- Geodesy at Ohio State University, Columbus OH, USA
- Geodesy and Geomatics Engineering, University of New Brunswick, Canada
- Faculty of Geodesy and Geomatics Engineering,K.N.Toosi University of Technology, Tehran,Iran
- Department of Geodesy and Geomatics, Zanjan University, Zanjan, Iran
- Department of Surveying and Geomatics Engineering at the University of Tehran,Tehran,Iran
- Department of Surveying at Helsinki University of Technology, Espoo, Finland
- Geomatics Engineering at the University of Calgary, Alberta, Canada
- Wuhan Technical University of Surveying and Mapping (WTUSM), Wuhan, China
- Department of Spatial Sciences, Curtin University of Technology, Perth, Australia
- Faculty of Geodesy and Geoinformatics, University of Zagreb, Zagreb, Croatia

- Faculty of Geoinformation Science & Engineering, Malaysian University of Technology, Johor Bahru, Malaysia
- Department of Geomatic Engineering, University Of Melbourne, Australia
- Escuela de Ingenieria Geodesica de La Universidad del Zulia *(LUZ-INGENIERIA GEODESICA) at Maracaibo, Venezuela*
- Geodesy division, Middle East Technical University Ankara, Turkey
- Department of Land Surveying and Geo-Informatics, LSGI, The Hong Kong Polytechnic University Hong Kong, SAR, China.

Governmental Agencies

- National Geodetic Survey (NGS) in Silver Spring MD, USA
- National Geospatial-Intelligence Agency (NGA) in Bethesda MD, USA (Previously National Imagery and Mapping Agency NIMA, previously Defense Mapping Agency DMA)
- U.S. Geological Survey (USGS) in Reston VA, USA
- Institut Géographique National (IGN) in Saint-Mandé, France
- Bundesamt für Kartographie und Geodäsie (BKG) In Frankfurt a.M., Germany (Previously Institut für Angewandte Geodäsie, IfAG)
- *Central Research Institute for Geodesy, Remote Sensing and Cartography* (CNIIGAIK), Moscow, Russia
- Geodetic Survey Division, Natural Resources Canada

- Geoscience Australia, Australian Federal Agency
- *Finnish Geodetic Institute* (FGI) at Masala, Finland
- *Portuguese Geographic Institute* (IGEO) at Lisbon, Portugal
- National Cartographic Center of Iran (NCC) Tehran, Iran
- National Geographic Organization of Iran (NGO) Tehran, Iran.

32

Climatology

Climatology is the study of climate, scientifically defined as weather conditions averaged over a period of time, and is a branch of the atmospheric sciences. In contrast to meteorology, which studies short term weather systems lasting up to a few weeks, climatology studies the frequency with which these weather systems occurred in the past. It studies the periodicity of weather events over years to millennia, as well as changes in long-term average weather patterns, in relation to atmospheric conditions. *Climatologists,* those who practice climatology, study both the nature of climates - local, regional or global - and the natural or human-induced factors that cause climates to change. Climatology considers the past and can help predict future climate change.

Phenomena of climatological interest include the atmospheric boundary layer, circulation patterns, heat transfer (radiative, convective and latent), interactions between the atmosphere and the oceans and land surface (particularly vegetation, land use and topography), and the chemical and physical composition of the atmosphere. Related disciplines include chemistry, ecology, geology, geophysics, glaciology, hydrology, oceanography, and volcanology.

Climatology is approached in a variety of ways. *Paleoclimatology* seeks to reconstruct past climates by examining records such as ice cores and tree rings (dendroclimatology). The study of contemporary climates incorporates meteorological data accumulated over many years, such as records of rainfall, temperature and atmospheric composition. Knowledge of the atmosphere and its dynamics is also embodied in models, either statistical or mathematical, which help by integrating different observations and testing how they fit together. Modeling is used for understanding past, present and potential future climates.

Climate research is made difficult by the large scale, long time periods, and complex processes which govern climate. It is generally accepted that climate is governed by differential equations based on physical laws, but what, exactly, are these equations, and what can be concluded from them, is still subject to debate. Climate is sometimes modeled as a stochastic process but this is generally accepted as an approximation to processes that are otherwise too complicated to analyze.

History of Climatology

Early climate researchers include Edmund Halley, who published a map of the trade winds in 1686, after a voyage to the southern hemisphere. Francis Galton invented the term *anticyclone*. Helmut Landsberg led to statistical analysis being used in climatology, which led to its evolution into a physical science.

Renowned Climatologists

- Francis Galton
- Edmund Halley
- Wladimir Köppen

- Helmut Landsberg
- Hubert Lamb
- Milutin Milankoviæ
- John Russell Mather

33

Palaeogeography

Palaeogeography (sometimes spelled palaeogeography) is the study of the ancient geologic environments of the Earth's surface as preserved in the stratigraphic record.

Paleogeographic analysis is used in the detailed study of sedimentary basins in petroleum geology. Peleogeographers also study the sedimentary environment associated with fossils to aid in the understanding of evolutionary development of extinct species. The reconstructions of prehistoric continents and oceans depends on paleogeographic evidence. Thus paleogeography provided critical evidence for the development of continental drift and current plate tectonic theories. For example, knowledge of the shape and latitudinal location of supercontinents such as Pangaea and ancient oceans such as Panthalassa result from paleogeographic studies.

- Plate tectonics: Behavior of plates covering the surface of the Earth.
- Paleontology: Study of ancient life, often involving fossils and pollen (palynology).
- Paleoclimatology: Study of ancient climates.

34

Paleoclimatology

Paleoclimatology is the study of climate change taken on the scale of the entire history of the Earth.

Techniques of Paleoclimatology

Paleoclimatologists employ a wide variety of skills to arrive at their theories and conclusions.

Glaciers are a widely employed instrument in paleoclimatology. The ice in glaciers has hardened into an identifiable pattern, with each year leaving a distinct layer in an ice core. It is estimated that the polar ice caps have 100,000 of these layers or more.

- Inside of these layers scientists have found pollen, allowing them to estimate the total amount of plant growth of that year by the pollen count. The thickness of the layer can help to determine the amount of rainfall that year.
- Because evaporation rates of water molecules with slightly heavier isotopes of hydrogen and oxygen are slightly different during warmer and colder periods, changes in the average temperature of the ocean surface are reflected in slightly different

ratios between those isotopes. Various cycles in those isotope ratios have been detected.

Petrified tree rings give paleoclimatology data over a much larger stretch of time. The fossil itself is dated with radioactive dating within a wide margin of error. The rings themselves can give some information about rainfall and temperature during that epoch.

Sediment layers have been studied, particularly those in the bottom of lakes and oceans. Characteristics of preserved vegetation, animals, pollen, and isotope ratios provide information.

Sedimentary rock layers provide a more compressed view of climate, as each layer of sediment is made over a period of hundreds of thousands to millions of years. Scientists can get a grasp of long term climate by studying sedimentary rock. Some scientists believe each layer designates a major change in climate.

Planet's Timeline

Some of the mile stones that mark the history of the planet are as follows (Ma = Millions of years ago):

4,000 Ma	earliest biogenic carbon
3,700 Ma	oldest rocks
3,500 Ma	oldest stromatolites
3,500 Ma	first evidence of sex [ref. *Origins of Sex : Three Billion Years of Genetic Recombination*]
3,450 Ma	earliest bacteria
3,800 Ma	banded iron formations (with reduced iron)
3,000 Ma	earliest precambrian ice ages [need ref]
[?]	Chuos Tillites of South-West Africa

[?]	Sturtian Tillites of the Finders Range, South-central Australia
3,000 Ma	earliest photosynthetic bacteria
2,700 Ma	oldest chemical evidence of complex cells
2,300 Ma	first green algae (eukaryotes)
2,000 Ma	free oxygen in the atmosphere
2,000 Ma to 1600 Ma	Gowganda tillites in the Canadian shield
1,700 Ma	end of the banded iron formations and red beds become abundant (non-reducing atmosphere)
700 Ma	first metazoans late Proterozoic (Ediacaran Epoch) - first skeletons
570 Ma to present	Phanerozic Eon [history of the phanerozoic goes in here]
100 Ma	development of the angiosperms (flowering plants)
2 Ma to present	modern world and man's appearance on earth [whole history of man goes in here about 1/2000th of the time scale]
0.01 Ma	end of the last ice age
0.001 Ma	warming trend of the middle ages
0.0001 Ma	end of the mini ice age
0.00022 Ma to present	industrialized human world.

Various Notable Climate-Related Events

- Periodic Ice ages
- Snowball Earth/Varangian glaciation (Hadean and Paleoproterozoic)
- Permian-Triassic extinction event (Permian-Triassic)
- Paleocene-Eocene Thermal Maximum (Paleocene-Eocene)
- Younger Dryas/The Big Freeze (~11 000 BCE)

- Holocene climatic optimum (~7000 BCE-3000 BCE)
- Climate changes of 535-536 (535-536)
- Medieval warm period (900-1300)
- Little ice age (1300-1800)
- Year Without a Summer (1816)

History of the Atmosphere

Earliest Atmosphere

The earliest atmosphere of the Earth was probably stripped away by solar winds early in the history of the planet. These gases were later replaced by an atmosphere derived from outgassing from the Earth. Sometime during the late Archean Era an oxygen atmosphere began to develop from photosynthesizing algae.

Carbon Dioxide and Free Oxygen

Free oxygen did not exist until about 1,700 Ma and this can be seen with the development of the red beds and the end of the banded iron formations. This signifies a shift from a reducing atmosphere to an oxidising atmosphere. The early atmosphere and hydrosphere (up until about 2,000 Ma) were devoid of free Oxygen. After photosynthesis developed, photoautotrophs began releasing O_2.

The very early atmosphere of the earth contained mostly carbon dioxide (CO_2) : about 80%. This gradually dropped to about 20% by 3,500 Ma. This coincides with the development of the first bacteria about 3,500 Ma. By the time of the development of photosynthesis (2,700 Ma), CO_2 levels in the atmosphere were in the range of 15%. During the period from about 2,700 Ma to about 2,000 Ma, photosynthesis dropped the CO_2 concentrations from about 15% to about 8%. By about 2,000 Ma free O_2 was beginning to accumulate. This gradual reduction in CO_2 levels continued to about 600 Ma at which point CO_2 levels were below 1% and O_2 levels had risen to more than 15%. 600Ma corresponds to the end of the Precambrian and the beginning of the Cambrian, the end of the cryptozoic and the beginning of the Phanerozic, and the beginning of oxygen-breathing life.

Precambrian Climate

The climate of the late Precambrian was typically cold with glaciation spreading over much of the earth. At this time the continents were bunched up in a supercontinent called Rodinia. Massive deposits of tillites are found and anomalous isotopic signatures are found which are consistent with the idea that the earth at this time was a massive snowball. Map of Rodinia at the end of the Precambrian after Australia and Antarctica rotated away from the southern hemisphere.

As the Proterozoic Eon drew to a close, the Earth started to warm up. By the dawn of the Cambrian and the Phanerozoic Eon, Earth was experiencing average global temperatures of about +22 °C. Hundreds of millions of years of ice were replaced with the balmy tropical seas of the Cambrian Period within which life exploded at a rate never seen before or after.

Phanerozoic Climate

Qualitatively, the Earth's climate was varied between conditions that support large-scale continental glaciation and those which are extensively tropical and lack permanent ice caps even at the poles. The time scale for this variation is roughly 140 million years and may be related to Earth's motion into and out of galactic spiral arms (Veizer and Shaviv 2003). The difference in global mean temperatures between a fully glacial earth and ice free Earth is estimated at approximately 10 °C, though far larger changes would be observed at high latitudes and smaller ones at low latitudes. One key requirement for the development of large scale ice sheets is the arrangement of continental land masses at or near the poles. With plate tectonics constantly rearranging the continents, it can also shape long-term climate evolution. However, the presence of land masses at the poles is not sufficient to guarantee glaciations. Evidence exists of past warm periods in Earth's climate when polar land masses similar to Antarctica were home to deciduous forests rather than ice sheets.

Changes in the atmosphere may also exert an important influence over climate change. The establishment of CO_2-consuming (and oxygen-producing) photosythesizing organisms in the Precambrian led to the production of an atmosphere much like today's, though for most of this period it was much higher in CO_2 than today. Similarly, the Earth's average temperature was also frequently higher than at present, though it has been argued that over very long time scales climate is largely decoupled from carbon dioxide variations (Veizer et al. 2000). Or more specifically that changing continental configurations and mountain building probably have a larger impact on climate than carbon dioxide. Others dispute this, and suggest that the variations of temperature in response to carbon dioxide changes have been underestimated (Royer et al. 2004). However, it is clear that the preindustrial atmosphere with only 280 ppm CO_2 is not far from the lowest ever occurring since the rise of macroscopic life.

Superimposed on the long-term evolution between hot and cold climates have been many short-term fluctuations in climate similar to, and sometimes more severe than, the varying glacial and interglacial states of the present ice age. Some of the most severe fluctuations, such as the Paleocene-Eocene Thermal Maximum, may be related to rapid increases in atmospheric carbon dioxide due to the collapse of natural methane reservoirs in the oceans (*see methane clathrates*). Severe climate changes also seem to have occurred during the course of the Cretaceous-Tertiary, Permian-Triassic, and Ordovician-Silurian extinction events; however, it is unclear to what degree these changes caused the extinctions rather than merely responding to other processes that may have been more directly responsible for the extinctions.

Quaternary Subera

The Quaternary subera includes the current climate. There has been a cycle of ice ages for the past 2.2-2.1 million years (starting before the Quaternary in the late Neogene Period).

Note in the graphic on the right the strong 120,000 year periodicity of the cycles, and the striking asymmetry of the curves. This asymmetry is believed to result from complex interactions of feedback mechanisms. It has been observed that ice ages deepen by progressive steps, but the recovery to interglacial conditions occurs in one big step.

Controlling Factors

Geologically short-term (<120,000 year) temperatures are believed to be driven by orbital factors (see Milankovitch cycles). The arrangements of land masses on the Earth's surface are believed to reinforce these orbital forcing effects.

Continental drift obviously affects the thermohaline circulation, which transfers heat between the equatorial regions and the poles, as does the extent of polar ice coverage.

The timing of ice ages throughout geologic history is in part controlled by the position of the continental plates on the surface of the Earth. When landmasses are concentrated near the polar regions, there is an increased chance for snow and ice to accumulate. Small changes in solar energy can tip the balance between summers in which the winter snow mass completely melts and summers in which the winter snow persists until the following winter. See the web site Paleomap Project for images of the polar landmass distributions through time.

Comparisons of plate tectonic continent reconstructions and paleoclimatic studies show that the Milankovitch cycles have the greatest effect during geologic eras when landmasses have been concentrated in polar regions, as is the case today. Today, Greenland, Antarctica, and the northern portions of Europe, Asia, and North America are situated such that a minor change in solar energy will tip the balance between year-round snow/ice preservation and complete summer melting. The presence of snow and ice is a well-understood positive feedback mechanism for climate. The Earth today is considered to be prone to ice age glaciations.

Another proposed factor in long term temperature change is the Uplift-Weathering Hypothesis, first put forward by T. C. Chamberlin in 1899 and later independently proposed in 1988 by Maureen Raymo and colleagues, where upthrusting mountain ranges expose minerals to weathering resulting in their chemical conversion to carbonates thereby removing CO_2 from the atmosphere and cooling the earth. Others have proposed similar effects due to changes in average water table levels and consequent changes in sub-surface biological activity and PH levels.

Over the very long term the energy output of the sun has gradually increased, on the order of 5% per billion (10^9) years, and will continue to do so until it reaches the end of its current phase of stellar evolution.

REFERENCES

Arthur Lovejoy (1936), *The Great Chain of Being*, Harvard University Press.

Athanasius Kircher (1665). *Mundus Subterraneus*, Amsterdam. See Marjorie Nicolson, *Mountain Gloom and Mountain Glory* (New York: Cornell University Press, 1959), pp. 168-73.

Bacon, 'De Dignitate et Augmentis Scientarum' in *Works,* ed. J. Spedding, R. Ellis. D. Heath (London: Longmans, 1870). vol. 4, p. 296.

Bradley, Raymond S. (1985.) *Quaternary paleoclimatology : methods of paleoclimatic reconstruction* (Boston: Allen & Unwin).

Buol, Stanley W., F.D. Hole and R.W. McCracken. 1997. Soil Genesis and Classification, 4th ed. Iowa State Univ. Press.

Carolyn Merchant, *The Death of Nature* (London: Wildwood, 1982), 'Organic Society and Utopia', pp. 69-99.

Crowley, Thomas J., and North, Gerald R. (1996) *Paleoclimatology* (Oxford, UK: Clarendon Press) Series : Oxford monographs on geology and geophysics no. 18.

Edwin Arthur Burtt, *The Metaphysical Foundations of Modern Physical Science (London:* Kegan Paul, 1925); ch. 3: 'Galileo', pp. 61-95.

Gould, Stephen Jay. (c. 1989) *Wonderful life, the story of the Burgess Shale* (New York: W.W. Norton).

Hole Francis D., J.B. Campbell. (1985). Soil landscape analysis. Totowa Rowman & Allanheld, p. 214.

Imbrie, John. (1986) *Ice ages : solving the mystery* (Cambridge, Massachusetts: Harvard University Press, 1986, c1979).

John Barrell, *The Dark Side of the Landscape: The Rural Poor in English Painting 1730-1840* (Cambridge University Press, 1980).

Lucile Brockway, *Science and Colonial Expansion: The Role of the British Royal Botanic Gardens* (London: Academic Press, 1979), pp. 141-65.

M.H. Abrams. *The Mirror and the Lamp* (Oxford University Press, 1953). pp. 298-335.

Margulis, Lynn, and Dorion Sagan. (c1986) *Origins of sex: three billion years of genetic recombination* (New Haven : Yale University Press) Series : The Bio-origins series.

Marjorie Nicolson, *Newton Demands the Muse* (Princeton University Press, 1946).

Merchant, *The Death of Nature;* Susan Griffin, *Woman and Nature* (New York: Harper and Row, 1978).

Miriam Rothschild, *Dear Lord Rothschild* (London: Hutchinson, 1983).

Niavis, Judicum Jovis (Leipzig, n.d.). For a fuller account of this allegory, see Frank Dawson Adams, *The Birth and Development of the Geological Sciences* (New York: Dover, 1938), pp. 171-5.

Pliny, *Natural History,* trans. J. Bostock and H. T. Riley (London: Bohn, 1858), vol. 6, Bk 33, ch. 1.

Ronald Amundsen. Soil Preservation and the Future of Pedology.

Royer, Dana L. and Robert A. Berner, Isabel P. Montañez, Neil J. Tabor, David J. Beerling (2004) "CO_2 as a primary driver of Phanerozoic climate", in *GSA Today* July 2004.

Shaviv, N. and Veizer, J. (2003) "Celestial driver of Phanerozoic climate?", in *GSA Today* July 2003, volume 13, number 7, pages 4-10.

Veizer, J., Godderis, Y. and François, L.M. (2000) "Evidence for decoupling of atmospheric CO_2 and global climate during the Phanerozoic eon", in *Nature*, volume 408, pages 698-701.

35

Geographic Coordinate System

A geographic coordinate system expresses every location on Earth by two of the three coordinates of a spherical coordinate system which is aligned with the spin axis of Earth. Borrowing from theories of ancient Babylonians, later expanded by the famous Greek thinker and geographer Ptolemy, a full circle is assigned 360 degrees (360°).

- latitude (Lat.) is the angle between any point and the equator. Lines of constant latitude are called parallels. They trace circles on the surface of Earth, but the only parallel that is a great circle is the equator (latitude=0 degrees), with each pole being 90 degrees (north pole 90° N; south pole 90° S).
- longitude (Long.) is the angle east or west of an arbitrary point on Earth: The Royal Observatory, Greenwich (UK) is the international zero-longitude point (longitude=0 degrees). The anti-meridian of Greenwich is both 180°W and 180°E. Lines of constant longitude are called meridians. The meridian passing through Greenwich is the Prime Meridian. Unlike parallels, all meridians are halves of great circles, and meridians are not parallel: they intersect at the north and south poles.

By combining these two angles, the horizontal position of any location on Earth can be specified.

For example, Baltimore, Maryland (in the USA) has a latitude of 39.3° North, and a longitude of 76.6° West (39.3° N 76.6° W). So, a vector drawn from the center of Earth to a point 39.3° north of the equator and 76.6° west of Greenwich will pass through Baltimore.

This latitude/longitude "webbing" is known as the common *graticule*. There is also a complementary transverse graticule (meaning the graticule is shifted 90°, so that the poles are on the horizontal equator), upon which all spherical trigonometry is ultimately based.

Traditionally, degrees have been divided into minutes (2) and seconds (3). There are several formats for degrees, all of them appearing in a Lat.-Long. order:

- DM Degree: Minute (49: 30.0-123: 30.0)
- DMS Degree: Minute: Second (49: 30: 00-123: 30: 00)
- DD Decimal Degree (49.5000-123.5000), generally with 4 decimal numbers.

To change from DM or DMS to DD, Decimal degrees = whole number of degrees, plus minutes divided by 60, plus seconds divided by 3600. Decimal division is now the most common and standard.

The equator is obviously an important part of this coordinate system, it represents the zeropoint of the latitude angle, and the halfway point between the poles. The equator is the fundamental plane of the geographic coordinate system. All spherical coordinate systems define such a fundamental plane.

Latitude and Longitude values are established based on an associated Geodetic system or datum such as WGS84. In other words, the same exact point on the earth's surface will be expressed by different latitude and longitude values, depending on the reference datum.

In popular GIS software, data projected in latitude/ longitude is often specified via a 'Geographic Coordinate System'. For example, data in latitude/longitude with the datum as the North American Datum of 1983 is denoted by 'GCS_North_American_1983'.

Geostationary Coordinates

Geostationary satellites (e.g., television satellites) are over the equator. So, their position related to Earth is expressed in longitude degrees. Their latitude does not change, and is always zero over the equator.

Third Dimension: Altitude, Height, Depth

To completely specify a location on, in, or above Earth, one has to also specify the elevation. The elevation specifies the vertical position of the location relative to some measure of Earth's surface. This could be expressed in terms of the vertical distance to Earth below, but, because of the ambiguity of "surface" and "vertical", is more commonly expressed relative to a more precisely defined datum such as mean sea level (as height above mean sea level) or the geoid. The distance to Earth's center is a practical coordinate both for very deep positions and for positions in space.

Various elevation / height coordinates either with respect to the surface or some other datum are altitude, height, and depth.

36

Landscape Ecology

Landscape ecology is a subdiscipline of ecology and geography that address how spatial variation in the landscape affects ecological processes such as the distribution and flow of energy, materials and individuals in the environment (which, in turn, may influence the distribution of landscape "elements" themselves such as hedgerows). Landscape ecology typically deals with problems in an applied and holistic context. The term landscape ecology was coined by Carl Troll, a German geographer in 1939 (Troll 1939). He developed this terminology and many early concepts of landscape ecology as part of his early work applying aerial photograph interpretation to studies of interactions between environment and vegetation.

One central landscape ecology theory originated from MacArthur & Wilson's *The Theory of Island Biogeography*. This work considered the assembly of flora and fauna on islands as the result of colonization from a mainland stock and stochastic extinction. The concepts of island biogeography were generalized from physical islands to abstract patches of habitat by Levins' metapopulation model. This generalization spurred the growth of landscape ecology by providing conservation biologists a new tool to assess how habitat fragmentation affects population viability. Reçent growth of landscape ecology owes much to the development of geographic information systems (GIS) technology and the availability of large-extent habitat data (e.g. remotely sensed satellite images or aerial photography).

Landscape ecology, a subdiscipline of ecology, addresses the causes and consequences of spatial heterogeneity (Forman 1995). Heterogeneity is the measure of how different parts of a landscape are from one another. Landscape ecology looks at how spatial structure affects organism abundance at the landscape level, as well as the behavior and functioning of the landscape as a whole. This includes the study of the pattern, or the internal order of a landscape, on process, or the continuous operation of functions of organisms (Turner 1989). Landscape ecology also includes geomorphology as applied to the design and architecture of landscapes (Allaby 1998). Geomorphology is the study of how geological formations are responsible for the structure of a landscape.

Development as a Discipline

The term "landscape ecology" is first attributed to the German geographer Carl Troll (1939). Landscape ecology developed in Europe from historical planning on human-dominated landscapes. In North America, concepts from general ecology theory were integrated. While general ecology theory and its sub-disciplines focused on the study of more homogenous, discrete community units organized in a hierarchical structure (typically as populations, species, and communities), landscape ecology built upon heterogeneity in space and time, and frequently included human-caused landscape changes in theory and application of concepts (Sanderson and Harris 2000),

By 1980, landscape ecology was a discrete, established discipline, marked by the organization of the International Association for Landscape Ecology (IALE) in 1982 and landmark book publications defining the scope and goals of the discipline, including Naveh and Lieberman (1984) and Forman and Godron (1986) (Ryszkowski 2002). Forman (1995) wrote that although study of "the ecology of spatial configuration at the human scale" was barely a decade old, there was strong potential for theory development and application of the conceptual framework. Today, theory and application of landscape ecology continues to develop through a need for innovative applications in a changing landscape and environment. Landscape ecology today relies more on advanced technologies such as remote sensing, GIS, and simulation models, with associated development of powerful quantitative methods to examine the interactions of patterns and processes (Turner et al. 2001). An example would be determining the amount of carbon present in the soil based on landform over a landscape, derived from GIS maps, vegetation types, and rainfall data for a region.

Relationship to Ecological Theory

Although landscape ecology theory may be slightly outside of the "classical and preferred domain of scientific disciplines" because of the large, heterogeneous areas of study (Sanderson and Harris 2000), general ecology theory is central to landscape ecology theory in many aspects. Landscape ecology is comprised of four main principles, which include: 1. the development and dynamics of spatial heterogeneity, 2. interactions and exchanges across heterogeneous landscapes, 3. influences of spatial heterogeneity on biotic and abiotic processes, and 4. the management of spatial heterogeneity. The main difference from traditional ecological studies, which frequently assume that systems are spatially homogenous, is the consideration of spatial patterns (Turner and Gardner 1991).

Important Terms in Landscape Ecology

Landscape ecology not only embraced a new vocabulary of terms but also incorporated general ecology theory terms in new ways. Many of the terms used in landscape ecology are as interconnected and interrelated as the discipline itself. Landscape can be defined as an area containing two or more ecosystems in close proximity (Sanderson and Harris 2000).

Scale and Heterogeneity (Incorporating Composition, Structure, and Function)

A main concept in landscape ecology is scale. Scale represents the real world as translated onto a map, in the relationship between distance on a map image and the corresponding distance on earth (Malczewski 1999). Scale is also the spatial or temporal measure of an object or a process (Turner and Gardner 1991), or level or degree of spatial resolution (Forman 1995). Components of scale include composition, structure, and function, which are all important ecological concepts. Applied to landscape ecology, composition refers to the number of patch types (see below) represented on a landscape, and their relative abundance. For example, the amount of forest or wetland, the length of forest edge, or the density of roads can be aspects of landscape composition. Structure is determined by the composition, the configuration, and the proportion of different patches across the landscape, while function refers to how each element in the landscape interacts based on its life cycle events (Turner and Gardner 1991). Pattern is the term for the contents and internal order of a heterogeneous area of land (Forman and Godron 1986).

A landscape with structure and pattern implies that it has spatial heterogeneity, or the uneven, non-random distribution of objects across the landscape (Forman 1995). Heterogeneity is a key element of landscape ecology that separates this discipline from other branches of ecology.

Patch and Mosaic

Patch, a term fundamental to landscape ecology, is defined as a relatively homogeneous area that differs from its surroundings (Forman 1995). Patches are the basic unit of the landscape that change and fluctuate, a process called patch dynamics. Patches have a definite shape and spatial configuration, and can be described compositionally by internal variables such as number of trees, number of tree species, height of trees, or other similar measurements (Forman 1995).

Matrix is the "background ecological system" of a landscape with a high degree of connectivity. Connectivity is the measure of how connected or spatially continuous a corridor, network, or matrix is (Forman 1995). For example, a forested landscape (the matrix) with fewer gaps in forest cover (open patches) will have higher connectivity. Corridors have important functions as strips of a particular type of landscape differing from adjacent land on both sides (Forman 1995). A network is an interconnected system of corridors while mosaic describes the pattern of patches, corridors and matrix that form a landscape in its entirety (Forman 1995).

Boundary and Edge

Landscape patches have a boundary between them which can be defined or fuzzy (Sanderson and Harris 2000). The zone composed of the edges of adjacent ecosystems is the boundary (Forman 1995). Edge means the portion of an ecosystem near its perimeter, where influences of the adjacent patches can cause an environmental difference between the interior of the patch and its edge. This edge effect includes a distinctive species composition or abundance in the outer part of the landscape patch (Forman 1995). For example, when a landscape is a mosaic of perceptibly different types, such as a forest adjacent to a grassland, the edge is the location where the two types adjoin. In a continuous landscape, such as a forest giving way to open woodland, the exact edge location is fuzzy and is sometimes determined by a local gradient exceeding a threshold, such as the point where the tree cover falls below thirty-five percent (Turner and Gardner 1991).

Ecotones, Ecoclines, and Ecotopes

A type of boundary is the ecotone, or the transitional zone between two communities (Allaby 1998). Ecotones can arise naturally, such as a lakeshore, or can be human-created, such as a cleared agricultural field from a forest (Allaby 1998). The ecotonal community retains characteristics of each bordering community and often contains species not found in the adjacent communities. Classic examples of ecotones include fencerows; forest to marshlands transitions; forest to grassland transitions; or land-water interfaces such as riparian zones in forests. Characteristics of ecotones include vegetational sharpness, physiognomic change, occurrence of a spatial community mosaic, many exotic species, ecotonal species, spatial mass effect, and species richness higher or lower than either side of the ecotone (Walker et al. 2003).

An ecocline is another type of landscape boundary, but it is a gradual and continuous change in environmental conditions of an ecosystem or community. Ecoclines help explain the distribution and diversity of organisms within a landscape because certain organisms survive better under certain conditions, which change along the ecocline. They contain heterogeneous communities which are considered more environmentally stable than those of ecotones (Attrill and Rundle 2002).

An ecotope is a spatial term representing the smallest ecologically-distinct unit in mapping and classification of landscapes (Forman 1995). Relatively homogeneous, they are spatially-explicit landscape units used to stratify landscapes into ecologically distinct features for measurement and mapping of landscape structure, function, and change over time, and to examine the effects of disturbance and fragmentation.

Disturbance and Fragmentation

Disturbance is an event that significantly alters the pattern of variation in the structure or function of a system, while fragmentation is the breaking up of a habitat, ecosystem, or land-use type into smaller parcels (Forman 1995). Disturbance is generally considered a natural process. Fragmentation causes land transformation, an important current process in landscapes as more and more development occurs.

Landscape Ecology Theory

Elements of Landscape Ecology Theory

Landscape ecology, as a theory, stresses the role of human impacts on landscape structures and functions and proposes ways for restoring degraded landscapes (Naveh and Lieberman 1984). Landscape ecology explicitly includes humans as entities that cause functional changes on the landscape (Sanderson and Harris 2000). Landscape ecology theory includes the landscape stability principle, which emphasizes the importance of landscape structural heterogeneity in developing resistance to disturbances, recovery from disturbances, and promoting total system stability (Forman and Godron 1986). This principle is a major contribution to general ecological theories which highlight the importance of relationships among the various components of the landscape. Integrity of landscape components helps maintain resistance to external threats, including development and land transformation by human activity (Turner et al. 2001). Analysis of land use changes has included a strongly geographical approach within landscape ecology. This has lead to acceptance of the idea of multifunctional properties of landscapes (Ryszkowski 2002). There are still calls for a more unified theory of landscape ecology due to differences in professional opinion among landscape ecologists, and the interdisciplinary approach to the discipline (Bastian 2001).

An important related theory is hierarchy theory, which refers to how systems of discrete functional elements operate when linked at two or more scales. For example, a forested landscape might be hierarchically composed of drainage basins, which in turn are composed of local ecosystems or stands, which are in turn composed of individual trees and tree gaps (Forman 1995). Recent theoretical developments in landscape ecology have emphasized the relationship between pattern and process, as well as the effect that changes in spatial scale has on the potential to extrapolate information across scales (Turner and Gardner 1991). Several studies suggest that the landscape has critical thresholds at which ecological processes will show dramatic changes, such as the complete transformation of a landscape by an invasive species with a small change in average temperatures per year which favors the invasive habitat requirements (Turner and Gardner 1991).

Landscape Ecology Application

Research Directions

Developments in landscape ecology illustrate the important relationships between spatial patterns and ecological processes, and incorporate quantitative methods that link spatial patterns and ecological processes at broad spatial and temporal scales. This linkage of time, space, and environmental change can assist land managers in applying land management plans to solve environmental problems (Turner et al. 2001). The increased attention in recent years on spatial dynamics has highlighted the need for new quantitative methods that can analyze patterns, determine the importance of spatially explicit processes on the landscape, and develop reliable landscape models (Turner and Gardner 1991). Multivariate analysis techniques, a type of statistics incorporating many variables, are frequently used to examine landscape level vegetation patterns. A number of studies in riparian systems and wetlands use a variety of statistical techniques, such as cluster analysis, canonical correspondence analysis (CCA), or detrended correspondence analysis (DCA), for classifying vegetation. Gradient analysis is another way to determine the vegetation structure across a landscape, or to help delineate critical wetland habitat for conservation or mitigation purposes (Lyon and Sagers 1998, Choesin and Boerner 2002).

Climate change is another major component in structuring current research in landscape ecology. Ecotones, as a basic unit in landscape studies, may have significance for management under climate change scenarios, since change effects are likely to be seen at ecotones first because of the unstable nature of a fringe habitat (Walker et al. 2003). Research in northern regions has examined landscape ecological processes, such as the accumulation of snow during winter, snow melting, freeze-thaw action, percolation, soil moisture variation, and temperature regimes through long-term measurements in Norway (Loffler and Finch 2005). The study analyzes gradients across space and time between ecosystems of the central high mountains to determine relationships between distribution patterns of animals in their environment. Looking at where animals live, and how vegetation shifts over time, may provide insight into changes in snow and ice over long periods of time across the landscape as a whole.

Other landscape-scale studies maintain that human impact is likely the main determinant of landscape pattern over much of the globe (Wilson and King 1995). Landscapes may become substitutes for biodiversity measures because plant and animal composition differs between samples taken from sites within different landscape categories. Taxa, or different species, can "leak" from one habitat into another, which has implications for landscape ecology. As human land use practices expand, and continue to increase the proportion of edges in landscapes, the effects of this leakage across edges on assemblage integrity may become more significant and important in conservation because taxa may be conserved across landscape levels, if not at local levels (Dangerfield et al. 2003).

Relationship to Other Disciplines

Landscape ecology has important links to application-oriented disciplines such as agriculture and forestry. In agriculture, landscape ecology has introduced new options for the control and management of environmental threats brought about by the intensification of agricultural practices. Agriculture has always been a strong human impact on ecosystems (Ryszkowski 2002). In forestry, changes in consumer needs have changed conservation and use of forested landscapes from structuring stands for fuelwood and timber to ordering stands across landscapes to enhance aesthetics, habitats and biological diversity. Landscape forestry provides methods, concepts, and analytic procedures for shifting management from traditional to landscape forestry (Boyce 1995). Landscape ecology has been cited as a major contributor to the development of fisheries biology as a distinct biological science discipline (Magnuson 1991), and is frequently incorporated in study design for wetland delineation in hydrology (Attrill and Rundle 2002).

- Biogeography
- Ecology
- Ecotope
- Schéma directeur d'aménagement et de gestion des eaux
- Total Human Ecosystem
- Historical Ecology

REFERENCES

Allaby, M. 1998. Oxford Dictionary of Ecology. Oxford University Press, New York, NY.

Attrill, M.J. and S.D. Rundle. 2002. Ecotone or ecocline: ecological boundaries in estuaries. Estuarine, Coastal, and Shelf Science 55: 929-936.

Boyce, S.G. 1995. Landscape Forestry. John Wiley and Sons, Inc., New York, NY.

Dangerfield, J.M., A.J. Pik, D.Britton, A. Holmes, M. Gillings, I. Oliver, D. Briscoe, and A.J. Beattie. 2003. Patterns of invertebrate biodiversity across a natural edge. Austral Ecology 28: 227-236.

Forman, R.T.T. and M. Godron. 1986. Landscape Ecology. John Wiley and Sons, Inc., New York, NY, USA.

Forman, R.T.T. 1995. Land Mosaics: The Ecology of Landscapes and Regions. Cambridge University Press, Cambridge, UK.

Loffler, J. and O.-D. Finch. 2005. Spatio-temporal gradients between high mountain ecosystems of central Norway. Arctic, Antarctic, and Alpine Research 37: 499-513.

Lyon, J. and C. L. Sagers, C.L. 1998. Structure of herbaceous plant assemblages in a forested riparian landscape. Plant Ecology 138: 1-16.

Magnuson, J.J. 1991. Fish and fisheries ecology. Ecological Applications 1: 13-26.

Malczewski, J. 1999. GIS and Multicriteria Decision Analysis. John Wiley and Sons, Inc., New York, NY, USA.

MacArthur, Robert H. and Wilson, Edward O. *The Theory of Island Biogeography* Princeton University Press. 2001 (reprint) ISBN 0-691-08836-5

Naveh, Z. and A. Lieberman. 1984. Landscape ecology: theory and application. Springer-Verlag, New York, NY, USA.

Ryszkowski, L. (ed.). 2002. Landscape Ecology in Agroecosystems Management. CRC Press, Boca Raton, Florida, USA.

Sanderson, J. and L.D. Harris (eds.). 2000. Landscape Ecology: A Top-Down Approach. Lewis Publishers, Boca Raton, Florida, USA.

Troll, C. 1939. Luftbildplan und ökologische Bodenforschung (Aerial photography and ecological studies of the earth). Zeitschrift der Gesellschaft für Erdkunde, Berlin: 241-298.

Turner, M.G. 1989. Landscape ecology: the effect of pattern on process. Annual Review of Ecology and Systematics 20: 171-197.

Turner, M.G. and R.H. Gardner (eds.). 1991. Quantitative Methods in Landscape Ecology. Springer-Verlag, New York, NY, USA.

Turner, M.G., R.H. Gardner and R.V. O'Neill, R.V. 2001. Landscape Ecology in Theory and Practice. Springer-Verlag, New York, NY, USA.

Walker, S., W.J. Barstow, J.B. Steel, G. L. Rapson, B. Smith, W. M. King, and Y. H. Cottam. 2003. Properties of ecotones: evidence from five ecotones objectively determined from a coastal vegetation gradient. Journal of Vegetation Science 14: 579-590.

Wilson, J.B. and W.M. King. 1995. Human-mediated vegetation switches as processes in landscape ecology. Landscape Ecology 10: 191-196.

37

Earthquake

An earthquake is a phenomenon that results from and is powered by the sudden release of stored energy that radiates seismic waves. At the Earth's surface, earthquakes may manifest themselves by a shaking or displacement of the ground and sometimes tsunamis, which may lead to loss of life and destruction of property.

Earthquakes may occur naturally or as a result of human activities. In its most generic sense, the word *earthquake* is used to describe any seismic event--whether a natural phenomenon or an event caused by humans--that generates seismic waves.

Types of Earthquakes

Naturally Occurring Earthquakes

Most naturally occurring earthquakes are related to the tectonic nature of the Earth. Such earthquakes are called *tectonic earthquakes*. The Earth's lithosphere is a patch work of plates (see plate tectonics) in slow but constant motion caused by the heat in the Earth's mantle and core. Plate boundaries glide past each other, creating frictional stress. When the frictional stress exceeds a critical value, called *local strength*, a sudden failure occurs. The boundary of tectonic plates along which failure occurs is called the *fault plane*. When the failure at the fault plane results in a violent displacement of the Earth's crust, the elastic strain energy is released and elastic waves are radiated, thus causing an earthquake. It is estimated that only 10 percent or less of an earthquake's total energy is ultimately radiated as seismic energy, while most of the earthquake's energy is used to power the earthquake fracture growth and is eventually converted into heat. Therefore, earthquakes lower the Earth's available potential energy and thermal energy, though these losses are negligible. To describe the physical process of occurrence of an earthquake, seismologists use the Elastic-rebound theory.

The majority of tectonic earthquakes originate at depths not exceeding a few tens of kilometers. Earthquakes occurring at boundaries of tectonic plates are called *interplate earthquakes*, while the less frequent events that occur in the interior of the lithospheric plates are called *intraplate earthquakes*.

Where the crust is thicker and colder, earthquakes occur at greater depths of hundreds of kilometers along subduction zones where plates descend into the Earth's mantle. These types of earthquakes are called *deep focus earthquakes*. They are possibly generated when subducted lithospheric material catastrophically undergoes a phase transition (e.g., olivine to spinel), releasing stored energy--such as elastic strain, chemical energy or gravitational energy--that cannot be supported at the pressures and temperatures present at such depths.

Earthquakes may also occur in volcanic regions and are caused by the movement of magma in volcanoes. Such quakes can be an early warning of volcanic eruptions.

A recently proposed theory suggests that some earthquakes may occur in a sort of earthquake storm, where one earthquake will trigger a series of earthquakes each triggered by the previous shifts on the fault lines, similar to aftershocks, but occurring years later, and with some of the later earthquakes as damaging as the early ones. Such a pattern was observed in the sequence of about a dozen earthquakes that struck the Anatolian Fault in Turkey in the 20th Century, the half dozen large earthquakes in New Madrid in 1811-1812, and has been inferred for older anomalous clusters of large earthquakes in the Middle East and in the Mojave Desert.

Induced Earthquakes

Some earthquakes have anthropogenic sources, such as extraction of minerals and fossil fuel from the Earth's crust, the removal or injection of fluids into the crust, reservoir-induced seismicity, massive explosions, and collapse of large buildings. Seismic events caused by human activity are referred to by the term *induced seismicity*. They however are not strictly earthquakes and usually show a different seismogram than earthquakes that occur naturally.

A rare few earthquakes have been associated with the build-up of large masses of water behind dams, such as the Kariba Dam in Zambia, Africa, and with the injection or extraction of fluids into the Earth's crust (e.g. at certain geothermal power plants and at the Rocky Mountain Arsenal). Such earthquakes occur because the strength of the Earth's crust can be modified by fluid pressure. Earthquakes have also been known to be caused by the removal of natural gas from subsurface deposits, for instance in the northern Netherlands. The world's largest reservoir-induced earthquake occurred on December 10, 1967 in the Koyna region of western Maharashtra in India. It had a magnitude of 6.3 on the Richter scale. However, the U.S. geological survey reported the magnitude of 6.8.

The detonation of powerful explosives, such as nuclear explosions, can cause low-magnitude ground shaking. Thus, the 50-megaton nuclear bomb code-named Ivan detonated by the Soviet Union in 1961 created a seismic event comparable to a magnitude 7 earthquake, producing the seismic shock so powerful that it was measurable even on its third passage around the Earth. In an effort to promote nuclear non-proliferation, the International Atomic Energy Agency uses the tools of seismology to detect illicit activities such as nuclear weapons tests. The nuclear nations routinely monitor each other's activities through networks of interconnected seismometers, which allow to precisely locate the source of an explosion.

Characteristics

Earthquakes occur on a daily basis around the world, most detected only by seismometers and causing no damage. Large earthquakes however can cause serious destruction and massive loss of life through a variety of agents of damage, including fault rupture, vibratory ground motion (shaking), inundation (tsunami, seiche, or dam failure), various kinds of permanent ground failure (liquefaction, landslides), and fire or a release of hazardous materials e.g gas leaks or petrol leaks. In a particular earthquake, any of these agents of damage can dominate, and historically each has caused major damage and great loss of life; nonetheless, for most earthquakes shaking is the dominant and most widespread cause of damage. There are four types of seismic waves that are all generated simultaneously and can be felt on the ground. Responsible for the shaking hazard, they are P-waves (primary waves), S-waves (secondary or shear waves) and two types of surfaces waves, (Love waves and Rayleigh waves).

Most large earthquakes are accompanied by other, smaller ones that can occur either before or after the main shock; these are called foreshocks and aftershocks, respectively. Aftershocks can be felt from half way round the world so in England you could feel an aftershock from New Zealand. While almost all earthquakes have aftershocks, foreshocks occur in only about 10% of events. The power of an earthquake is always distributed over a significant area, but in large earthquakes, it can even spread over the entire planet. Ground motions caused by very distant earthquakes are called teleseisms. The Rayleigh waves from the Sumatra-Andaman Earthquake of 2004 caused ground motion of over 1 cm even at seismometers that were located far from it, although this displacement was abnormally large. Using such ground motion records from around the world, seismologists can identify a point from which the earthquake's seismic waves apparently originated. That point is called its focus or hypocenter and usually coincides with the point where the fault slip started. The location on the surface directly above the hypocenter is known as the epicenter. The total length of the section of a fault that slips, the rupture zone, can be as long as 1,000 km for the biggest earthquakes.

Earthquakes that occur below sea level and have large vertical displacements can give rise to tsunamis, either as a direct result of the deformation of the sea bed due to the earthquake or as a result of submarine landslides directly or indirectly triggered by the quake.

Measuring Earthquakes

Since seismologists cannot directly observe rupture in the Earth's interior, they rely on geodetic measurements and numerical experiments to analyze seismic waves and accurately assess severity of earthquakes. Such analyses allow scientists to estimate the locations and likelihoods of future earthquakes, helping identify areas of greatest hazard and ensure safety of people and infrastructure located in such areas.

Severity

The severity of an earthquake is described by both magnitude and intensity. These two frequently-confused terms both refer to different, but related, observations. *Magnitude,* usually expressed as an Arabic numeral, characterizes the size of an earthquake by measuring indirectly the energy released. By contrast, *intensity* indicates the local effects and potential for damage produced by an earthquake on the Earth's surface as it affects humans, animals, structures, and natural objects such as bodies of water. Intensities are usually expressed in roman numerals, each representing the severity of the shaking resulting from an earthquake. Any given earthquake can be described by only one *magnitude*, but many *intensities* since the earthquake effects vary with circumstances such as distance from the epicenter and local soil conditions.

Charles Richter, the creator of the Richter magnitude scale, distinguished *intensity* and *magnitude* as follows: "I like to use the analogy with radio transmissions. It applies in seismology because seismographs, or the receivers, record the waves of elastic disturbance, or radio waves, that are radiated from the earthquake source, or the broadcasting station. Magnitude can be compared to the power output in kilowatts of a broadcasting station. Local intensity on the Mercalli scale is then comparable to the signal strength on a receiver at a given locality; in effect, the quality of the signal. Intensity, like signal strength, will generally fall off with distance from the source, although it also depends on the local conditions and the pathway from the source to the point."

Two fundamentally different but equally important types of scales are commonly used by seismologists to describe earthquakes. The original force or energy of an earthquake is measured on a *magnitude scale*, while the intensity of shaking occurring at any given point on the Earth's surface is measured on an *intensity scale*.

Seismic Intensity Scales

The first simple classification of earthquake intensity was devised by Domenico Pignataro in the 1780s. However, the first recognisable intensity scale in the modern sense of the word was drawn up by P.N.G. Egen in 1828; it was ahead of its time. The first widely adopted intensity scale, the Rossi-Forel scale, was introduced in the late 19th century. Since then numerous intensity scales have been developed and are used in different parts of the world: the scale currently used in the United States is the Modified Mercalli scale (MM), while the European Macroseismic Scale is used in Europe, the Shindo scale is used in Japan, and the MSK-64 scale is used in India, Israel, Russia and throughout the CIS. Most of these scales have twelve degrees of intensity, which are roughly equivalent to one another in values but vary in the degree of sophistication employed in their formulation.

Magnitude Scales

The first attempt to qualitatively define a single, absolute value to describe the size of earthquakes was the magnitude scale (the name being taking from similarly formulated scales used to represent the brightness of stars).

The local magnitude scale (M_L), also popularly known as the Richter Scale, is a quantitative logarithmic scale. In the 1930s, California seismologist Charles F. Richter devised a simple numerical scale to describe the relative sizes of earthquakes in Southern California. The name "Richter Scale" was coined by journalists and is not generally used by seismologists in technical literature. M_L is obtained by measuring the maximum amplitude of a recording on a Wood-Anderson torsion seismometer (or one calibrated to it) at a distance of 600 km from the earthquake. Other more recent magnitude measurements include: body wave magnitude (m_b), surface wave magnitude (M_s), and duration magnitude (M_D). Each of these is scaled to give values similar to those given by the local magnitude scale; but because each is based on a measurement of one part of the seismogram, they do not measure the overall power of the source and can be negatively affected by saturation at higher magnitude values--meaning that they fail to report higher magnitude values for larger events. This problem sets in at around magnitude 6 for local magnitude; surface-wave magnitude saturates above 8. Despite the limitations of older magnitude scales, they are still in wide use, as they can be calculated rapidly, catalogues of them dating back many years are available, and the public is familiar with them.

Because of the limitations of the magnitude scales, a new, more uniformly applicable extension of them, known as moment magnitude (M_W), was developed. In particular, for very large earthquakes moment magnitude gives the most reliable estimate of earthquake size. This is because seismic moment is derived from the concept of moment in physics and therefore provides clues to the physical size of an earthquake--the size of fault rupture and accompanying displacement and length of slippage--as of as well as the amount of energy released. So while seismic moment, too, is calculated from seismograms, it can also be obtained by working backwards from geologic estimates of the size of the fault rupture and displacement. The values of moments for different earthquakes range over several orders of magnitude, and because they are not influenced by variables such as local circumstances, the results obtained make it easy to objectively compare the sizes of different earthquakes. These characteristics, plus the seismic moment's immunity to saturation at higher magnitudes and compatibility with other magnitude scales, led Tom Hanks and Hiroo Kanamori to introduce in 1979 the moment magnitude (M_W) scale for representing the absolute size of earthquakes.

Seismic Maps

To show the extent of various levels of seismic effects within a particular locality, seismologists compile special maps called isoseismal maps. An isoseismal map uses contours to outline areas of equal value in terms of ground shaking intensity, ground surface liquefaction, shaking amplification, or other seismic effects. Typically, these maps are created by combining historical instrument-recorded data with responses to postal questionnaires that are sent to each post office near the earthquake and to a sparser sample of post offices with increasing distance from the earthquake. This way of preparing a seismic hazard map can take months to complete. In contrast to the old method, a newer method of information collection takes advantage of the Internet to generate initial hazard maps almost instantly. Data are received through a questionnaire on the Internet answered by people who actually experienced the earthquake, reducing the process of preparing and distributing a map for a particular earthquake from months to minutes.

Seismic hazard maps have many applications. They are used by insurance companies to set insurance rates for properties located in earthquake-risky areas, by civil engineers to estimate the stability of hillsides, by organizations responsible for the safety of nuclear waste disposal facilities, and also by building codes developers as the basis of design requirements.

In building codes, the shaking-hazard maps are converted into seismic zone maps, which are used for seismic analysis of structural components of buildings. The seismic zone maps depict seismic hazards as zones of different risk levels. Such zones are typically designated as Seismic Zone 0, Seismic Zone 1, Seismic Zone 2 and so on. The seismic zone maps usually show the severity of expected earthquake shaking for a particular level of probability, such as the levels of shaking that have a 1-in-10 chance of being exceeded in a 50-year period. Buildings and other structures must be designed with adequate strength to withstand the effects of probable seismic ground motions within the Seismic Zone where the building or structure is being constructed.

Size and Frequency of Occurrence

Small earthquakes occur every day all around the world, and often multiple times a day in places like California and Alaska in the U.S., as well as Indonesia and Japan on the other side of the Pacific. Large earthquakes occur less frequently, the relationship being exponential; namely, roughly ten times as many earthquakes larger than magnitude 4 occur in a particular time period than earthquakes larger than magnitude 5. For example, it has been calculated that the average recurrence for the United Kingdom can be described as follows:

- an earthquake of 3.7 or larger every year
- an earthquake of 4.7 or larger every 10 years
- an earthquake of 5.6 or larger every 100 years.

Most of the world's earthquakes (90%, and 81% of the largest) take place in the 40,000 km-long, horseshoe-shaped zone called the circum-Pacific seismic belt, also known as the Pacific Ring of Fire, which for the most part bounds the Pacific Plate.[3][4] Massive earthquakes tend to occur along other plate boundaries, too, such as along the Himalaya Mountains.

Effects/Impacts of Earthquakes

There are many effects of earthquakes, these include, but are not limited to,

- Broken windows, which could seriously injure anyone walking along a sidewalk
- Collapse of buildings, causing occupants to be trapped and causing deaths as such
- Fires, as seen in the 1906 San Francisco earthquake
- Tsunamis, as seen in the 2004 Sumatran earthquake
- Landslides
- Destabilisation of the base of some buildings which may lead to collapse in a future earthquake
- Disease
- Lack of basic necessities

Preparation for Earthquakes

- Emergency preparedness

- Household seismic safety
- Seismic retrofit
- Earthquake prediction

Specific Fault Articles

- Alpine Fault
- Calaveras Fault
- Hayward Fault Zone
- North Anatolian Fault Zone
- New Madrid Fault Zone
- San Andreas Fault
- Great Sumatran fault

Specific Earthquake Articles

Pre-20th Century

- Shaanxi Earthquake (1556). Deadliest known earthquake in history, estimated to have killed 830,000 in China.
- Dover Straits earthquake of 1580 (1580)
- Cascadia Earthquake (1700).
- Kamchatka earthquakes (1737 and 1952).
- Lisbon earthquake (1755).
- New Madrid Earthquake (1811).
- Fort Tejon Earthquake (1857).
- Charleston earthquake (1886). Largest earthquake in the southeastern United States, killed 100.
- Assam earthquake of 1897 (1897). Large earthquake that destroyed all masonry structures, measuring more than 8 on the Richter scale.

20th Century

- San Francisco Earthquake (1906). Between 7.7 and 8.3 magnitudes; killed approximately 3,000 people and caused around $400 million in damage
- Great Kanto earthquake (1923). On the Japanese island of Honshu, killing over 140,000 in Tokyo and environs.
- Napier earthquake (1931). 7.8 on the Richter scale; 256 dead.
- Assam earthquake of 1950 (1950). Earthquake in Assam measures 8.6M.
- Kamchatka earthquakes (1952 and 1737).
- Quake Lake (1959) 7.5 on Richter scale. Formed a lake in southern Montana, USA
- Great Chilean Earthquake (1960). Biggest earthquake ever recorded, 9.5 on Moment magnitude scale.
- Good Friday Earthquake (1964) In Alaska, it was the third biggest earthquake recorded, measuring 9.2M.
- Ancash earthquake (1970). Caused a landslide that buried the town of Yungay, Peru; killed over 40,000 people.
- Sylmar earthquake (1971). Caused great and unexpected destruction of freeway bridges and flyways in the San Fernando Valley, leading to the first major seismic retrofits of these types of structures, but not at a sufficient pace to avoid the next California freeway collapse in 1989.
- Tangshan earthquake (1976). The most destructive earthquake of modern times. The official death toll

was 255,000, but many experts believe that two or three times that number died.

- Guatemala (1976). 7.5 on the Richter Scale, causing 23,000 deaths, 77,000 injuries and the destruction of more than 250,000 homes.
- Great Mexican Earthquake (1985). 8.1 on the Richter Scale, killed over 6,500 people (though it is believed as many as 30,000 may have died, due to missing people never reappearing.)
- Whittier Narrows earthquake (1987).
- Armenian earthquake (1988). Killed over 25,000.
- Loma Prieta earthquake (1989). Severely affecting Santa Cruz, San Francisco and Oakland in California. This is also called the World Series Earthquake. It struck as the World Series was just getting underway. Revealed necessity of accelerated seismic retrofit of road and bridge structures.
- Northridge, California earthquake (1994). Damage showed seismic resistance deficiencies in modern low-rise apartment construction.
- Great Hanshin earthquake (1995). Killed over 6,400 people in and around Kobe, Japan.
- Ýzmit earthquake (1999) Killed over 17,000 in northwestern Turkey.
- Düzce earthquake (1999)
- Chi-Chi earthquake (1999)
- Baku earthquake (2000).

21st Century

- Nisqually Earthquake (2001).
- Gujarat Earthquake (2001).

- Dudley Earthquake (2002).
- Bam Earthquake (2003). Over 40,000 people are reported dead.
- Parkfield, California earthquake (2004). Not large (6.0), but the most anticipated and intensely instrumented earthquake ever recorded and likely to offer insights into predicting future earthquakes elsewhere on similar slip-strike fault structures.
- Chuetsu Earthquake (2004).
- Indian Ocean Earthquake (2004). One of the largest earthquakes in recorded history, registering a moment magnitude of 9.1-9.3. Epicentered off the coast of the Indonesian island of Sumatra, this massive tremor triggered a series of gigantic tsunamies that smashed onto the shores of a number of nations, causing more than 229,000 fatalities. It is regarded as one of the worst natural disasters ever to have hit the planet.
- Sumatran Earthquake (2005).
- Fukuoka Earthquake (2005).
- Kashmir Earthquake (2005). Killed over 79,000 people. Many more at risk from the Kashmiri winter. -update needed.
- Lake Tanganyika Earthquake (2005).
- May 2006 Java Earthquake (2006).
- July 2006 Java Earthquake (2006).

REFERENCE

Hiroo Kanamori, Emily E. Brodsky (April 2001). "The Physics of Earthquakes". *Physics Today.*

Appendix

WORLD RENOWNED GEOGRAPHERS

The geographers are listed below in English alphabetical transliteration order (by surnames).

- Abulfeda (Syria, 1273 - 1331)
- Agatharchides
- Agathedaemon of Alexandria
- Johnathon Agnew
- Stuart Aitken
- Alypius of Antioch (Roman Empire, fl. circa 450)
- Hafiz-i Abru (Persia, 15th century)
- Ash Amin
- Karl Andree (Germany, 1808 - 1875)
- Ibn Batuta (Morocco, 1304 - 1369)
- Brian J.L. Berry
- Al bruni
- William R. Black
- Nick Blomley
- Alison Blunt
- Bruce Braun
- William Bunge

- Anne Buttimer
- David Campbell
- Manuel Castells
- Walter Christaller (Germany 1893 - 1969)
- Paul Cloke
- Denis Cosgrove (England 1948)
- Kevin Cox
- Susan Cutter
- Mike Crang
- Phill Crang
- Jovan Cvijiæ (Serbia 1865 - 1927)
- William Morris Davis (USA, 1850 - 1934)
- John Dee (England, 1527 - 1608)
- Jared Diamond (USA, 1937)
- Dicaearchus (Greece, circa 350 BC - circa 285 BC)
- Eratosthenes (Greece, 276 BC - 194 BC)
- Stuart Elden
- Colin Flint
- Artur Gavazzi (Croatia, Slovenia, 1861 - 1944)
- Reginald G. Golledge
- Michael Frank Goodchild (1944 -)
- Jean Gottman
- Derek Gregory
- Pyrs Gruffud
- Torsten Hägerstrand (Sweden, 1916-2004)
- Richard Hartshorne (United States, 1899-1992)
- Karl Haushoffer

- David Harvey (United States, Britain, b. 1935)
- Gamal Hemdan (Egypt, 1928-1993)
- Hipparchus (Greece, 190 BC - 120 BC)
- Alexander von Humboldt (Germany, 1769 - 1859)
- Johann Homann (Germany, 1664 - 1724)
- Michael Jordan (United States - well-known basketball star, received bachelor's degree in 1986)
- John Paul Jones
- Peter Kropotkin
- Robert Kruse
- Andrew Latham
- Philippe Le Billon
- David Ley
- Michal Luknis (Slovakia, 1916-1986)
- Kenneth Mason, (UK 1887 - 1976)
- Doreen Massey
- John May
- Halford John Mackinder, (UK 1861 - 1947)
- Emil Mazur (Slovakia, 1925-1990)
- Linda McDowell
- Megasthenes
- August Meitzen, (Germany 1822 - 1910)
- Anton Melik (Slovenia, 1890 - 1966)
- Don Mitchell
- Brent W. Moll
- Kenneth Olwig
- Gearóid Ó Tuathail

- Orlando Ribeiro
- David Baines
- Pausanias (Greece, fl. circa 150)
- Allan Pred
- Pseudo-Scymnus
- Posidonius (Greece 135 BCE - 51 BCE)
- Ptolemy (Ptolemaic Egypt, Greece, circa 85 - circa 165)
- Pytheas (Greece, fl. circa 360 BC)
- Amin Razi (Persia, 16th century).
- Elisee Reclus
- Ferdinand Baron Von Richthofen 19th Century
- Barry Riddell
- Carl Ritter (Germany, 1779 - 1859.
- Ibn Rustah (Persia, d. 903)
- Arun Saldanha
- Marwyn Samuels
- Edward Said
- Carl Sauer (USA, 1889 -- 1975)
- Scymnus
- Andrew Shears
- Eric Sheppard
- Neil Smith
- Strabo (Roman Empire, Greece, 63 BC/64 BC - circa 24 BC)
- Edward Soja (Spain, 1942)
- Simon Springer

- Ulf Strohmayer
- Milton Santos
- Johann Heinrich von Thünen (Germany, 1783 - 1850)
- Nigel Thrift
- Waldo R. Tobler
- Yi-Fu Tuan (United States, China, b. 1930)
- James A. Tyner
- Paul Vidal de la Blache
- Arie van Deursen (Holland, 1891 - 1963)
- James E. Vance
- Alfred Weber (Germany, 1868 - 1958)
- Denis Wood (United States, 1945 -)
- Martin Zeiller (Germany, 1589 - 1661)
- Wilbur Zelinsky (United States, B. 1921)

Glossary of Terms

Abiotic: Non-living thing. Usually refers to the physical and chemical components of an organism's environment. Also called inorganic.

Ablation: Surface removal of ice or snow from a glacier or snowfield by melting, sublimation, and/or calving.

Ablation Zone: Region in a eer vapor in a given volume of air (this measurement is not influenced by the mass of the air). Normally expressed in grams of water vapor per cubic meter of atmosphere at a specific temperature.

Absolute Zero: Temperature of -273.15° Celsius. At this temperature atomic motion stops.

Abstract Space: Geographic model or representation of the real world. For example, maps and globes are abstractions of the real world or concrete space.

Abyssal Fan: Fan shaped accumulation of sediment from rivers that is deposited at the base of a submarine canyon within a ocean basin.

Abyssal Plain: Another name for ocean floor.

Acclimation: Slow adjustment of an organism to new conditions in its environment.

Accretion: The growth of the continental masses over geologic time via the addition of marine sediments. These sediments are added on to the edges of the continents through tectonic collision with other oceanic or continental plates.

Accumulation: Surface addition of snow to a glacier or snowfield.

Acidic: Any substance with a pH below 7.

Acid Precipitation: Atmospheric precipitation with a pH less than 5.6. Normal pH of precipitation is 5.6.

Acid Rain: Rain with a pH less than 5.6. Normal pH of precipitation is 5.6.

Acid Shock: A sudden acidification of runoff waters from the spring melting of accumulated snow in the middle latitudes because of the winter deposition of acidic precipitation.

Actinomycetes: Group of filamentous microorganisms that are intermediate between bacteria and fungi.

Active Layer: Upper zone of soil in higher latitude locations that experiences daily and seasonal freeze-thaw cycles.

Active Remote Sensing: Form of remote sensing where the sensor provides its own source of electromagnetic radiation to illuminate the object understudy. Radar is an example of an active remote sensing device.

Actual Evapotranspiration: Is the amount of water that is actually removed from a surface due to the processes of evaporation and transpiration.

Actual Mixing Ratio: Another term used to describe mixing ratio.

Adiabatic Cooling: The cooling of a rising parcel of air due to adiabatic processes.

Advection: Advection involves the transfer of heat energy by means of horizontal mass motions through a medium.

Aeolian: Geomorphic process involving wind. Alternative spelling *eolian*.

Aeolian Landform: Is a landform formed from the erosion or deposition of weathered surface materials by wind. This includes landforms with some of the following geomorphic features: sand dunes, deflation hollows, and desert pavement. Alternative spelling *eolian landform*.

Aerial Photography: Form of remote sensing that captures images of objects using photographic cameras and film from platforms in the atmosphere.

Aftershock: Smaller earth tremors that occur seconds to weeks after a major earthquake event.

Aggradation: Readjustment of the stream profile where the stream channel is raised by the deposition of bed load.

Agronomy: Field of science that studies phenomena related to agriculture.

Air Pollution: Toxification of the atmosphere through the addition of one or more harmful substances in the air. Substance must be in concentrations high enough to be hazardous to humans, other animals, vegetation, or materials. Also see primary pollutant and secondary pollutant.

Aleutian Low: Subpolar low pressure system found near the Aleutian Islands. Most developed during the winter season. This large-scale pressure system spawns mid-latitude cyclones.

Alfisols: Soil order (type) of the United States Natural Resources Conservation Service Soil Classification System. Soil associated with forest vegetation. Upper layers of this soil are relatively rich in organic matter. Whitish layer found in the A horizon because of eluviation. Illuvial layer forms in the B horizon.

Algae: A simple photosynthetic plant that usually lives in moist or aquatic environments. The bodies of algae can be unicellular or multicellular is design.

Alien Species: Species that is not naturally found in a region.

Allele: Alternative forms of a gene. Each form produces a unique inheritable characteristic.

Allelopathy: A particular form of amensalism found in plants. In this interaction, one species produces and releases of chemical substances that inhibit the growth of another species.

Allogenic Succession: A succession caused by a change in environmental conditions that is unrelated to the activities of the developing plant community.

Allopatric Speciation: The evolution of a new species because of the isolation of a small group of individuals from the other members of a population.

Allophane: Hydrated aluminosilicate substance ordinarily found associated with clay minerals.

Alluvial Fan: Large fan shaped terrestrial deposit of alluvial sediment on which a braided stream flows over. Form as stream load is deposited because of a reduction in the velocity of stream flow.

Alluvial Terraces: Flat elevated benches composed of unconsolidated alluvium found either side of a stream channel. Formed when a stream down cuts into its floodplain.

Alluvium: Sediment that originates from a stream.

Alpha Particle: Particle of matter that is positively charged. This particle consists of two neutrons and two protons and is emitted as a form of radioactivity from the nuclei of some radioisotopes.

Alpine Glacier: Small glacier that occupies a U-shaped valley on a mountain. Also called a *mountain glacier*.

Alpine Permafrost: Form of permafrost that exists at high altitudes in mountainous environments.

Alpine Tundra: High altitude biome dominated by a few species of dwarf shrubs, a few grasses, sedges, lichens, and mosses. Productivity is low in this biome because of the extremes of climate. Quite similar to tundra.

Alternative Hypothesis (H1): Is a hypothesis that has been suggested because it is believed to be false or because it is to be used as a starting point for scien 06/09/2006 9: 00 ng to organize arguments.

Altitude: Vertical distance above sea-level.

Altocumulus Clouds: Middle altitude cloud that is colored from white to gray. This cloud is composed of a mixture of water droplets and ice crystals. It appears in the atmosphere as layers or patches that are well rounded and commonly wavelike. Found in an altitude range from 2,000 to 8,000 meters.

Altostratus Clouds: Gray-looking middle altitude cloud that is composed of water droplets and ice crystals. Appears in the atmosphere as dense sheet like layer. Can be recognized from stratus clouds by the fact that you can see the sun through it. Found in an altitude range from 2,000 to 8,000 meters.

Amensalism: Interspecific interaction where one species suffers in terms of fitness, while the fitness of the other species does not change.

Amino Acid: Organic nitrogen containing acids which are used to construct proteins.

Ammonia: Chemical compound composed of nitrogen and hydrogen (NH_3). Component of the nitrogen cycle. Immediately released from organic matter upon decomposition.

Ammonium: Chemical compound composed of nitrogen and hydrogen (NH_4). Component of the nitrogen cycle. Product of organic matter decomposition. Can be fixed to clay minerals and later exchanged.

Amphibian: Group of vertebrate animals that can inhabit both terrestrial and aquatic habitats. This group of animals consists of frogs, newts, and salamanders. These organisms live at the land/water interface and spend most of their life in water. Descended from fish and ancestors to reptiles.

Amphibole: A group of double chained inosilicate minerals whose basic chemical unit is the tetrahedron (SiO_4). They are common rock forming minerals and are found in most igneous and metamorphic rocks. They form at low temperatures with the presence of water in the crystallization environment. There are about 60 recognized mineral types in this group.

Andesite: An extrusive igneous rock that develops from a magma that is chemically between felsic and mafic and whose mineral crystals are fine.

Aquiclude: Rock formations that are impermeable to groundwater water.

Aquifer Recharge Area: Surface area that provides water for an aquifer.

Archaea: Is a group of recently discovered organisms that resemble bacteria. However, these organisms are biochemically and genetically very different from bacteria. Some species of the domain Archaea live in the most extreme environments found on the Earth.

Archaebacteria: Term used to describe organisms that belong to the biological domain Archaea.

Archean Geologic eon that occurred from 2500 to 3800 million years ago. During this time period, the first single-celled prokaryote organisms evolved and developed.

Archipelago: A group of islands that have an arc shaped distribution. These islands are usually of volcanic origin and are associated with subduction zones.

Area Studies Tradition: Academic tradition in modern Geography that investigates an area on the Earth from a geographic perspective at either the local, regional, or global scale.

Arête: Sharp topographic ridge that separates cirques on a mountain that is or has been glaciated.

Arkose: A type of sedimentary sandstone that contains a large quantity of weathered feldspar grains. This type of sedimentary rock forms in arid conditions.

Aridisols: Soil order (type) of the United States Natural Resources Conservation Service Soil Classification System. Aridisols are commonly found in dry environments that are low in organic matter and rich in deposited salts.

Artesian Water: Groundwater that is confined by two impermeable layers beneath the Earth's surface.

Artesian Well: A well where the water rises and flows out to the surface because of hydrostatic pressure.

Arctic Circle: Latitude of 66.5° North. The southern limit of the area of the Earth that experiences 24 hours of darkness or 24 hours of day at least one day during the year.

Asexual Reproduction: Any process of reproduction that does not involve the fusion of gametes.

Asthenosphere: Zone in the Earth's mantle that exhibits

plastic properties. Located below the lithosphere at between 100 and 200 kilometers.

Astronomy: Field of knowledge that studies the nature, motion, origin, and constitution of celestial bodies.

Atmosphere: The atmosphere is the vast gaseous envelope of air that surrounds the Earth. Its boundaries are not easily defined. The atmosphere contains a complex system of gases and suspended particles that behave in many ways like fluids. Many of its constituents are derived from the Earth by way of chemical and biochemical reactions.

Atmospheric Pressure: Weight of the atmosphere on a surface. At sea-level, the average atmospheric pressure is 1013.25 millibars. Pressure is measured by a device called a barometer.

Atmospheric Stability: Relative stability of parcels of air relative to the atmosphere that surrounds them. Three conditions are generally described: stable, unstable, and neutral.

Atoll: A ring shaped reef composed largely of coral. These features are quite common in the tropical waters of the Pacific Ocean.

Atom: Smallest unit of an element that still maintains its chemical characteristics.

Atomic Energy: Energy released from an atomic nucleus at the expense of its mass.

Atomic Mass Number: Combined number of an atom's protons and neutrons.

Atomic Number: Number of protons in the nucleus of an atom.

Atomic Weight: Combined weight of an atom's electrons, protons, and neutrons.

Aurora: Multicolored lights that appear in the upper atmosphere (ionosphere) over the polar regions and visible from locations in the middle and high latitudes. Caused by the interaction of solar wind with oxygen and nitrogen gas in the atmosphere. Aurora in the Northern Hemisphere are called *aurora borelis* and *aurora australis* in the Southern Hemisphere.

Autogenic Succession: Succession where the plant community causes the environment to change and this modification drives the succession.

Autumnal Equinox: One of the two periods when the declination of the sun is at the equator. The autumnal equinox occurs on September 22 or 23.

Available Water: Portion of the capillary water that is available for plant root uptake.

Average Global Temperature: Average annual temperature of the Earth's entire surface atmosphere.

Azimuth: A system that measures direction clockwise from North over 360°.

Azonal soil: A soil without developed horizons.

Backscattering: Portion of solar radiation directed back into space as a result of particle scattering in the atmosphere.

Backshore: Area behind the shore. This coastal feature is located between the beach berm and the backshore slope.

Backshore slope: Sloping bank landward of the shore. This coastal feature is composed of relatively non-mobile sediments.

Backswamp: Marshy low lying area in a stream's floodplain. Commonly found behind levees.

Backwash: The return water flow of swash. This sheet of water flows back to ocean because of gravity.

Bacteria: Simple single celled prokaryotic organisms. Many different species of bacteria exist. Some species of bacteria can be pathogenic causing disease in larger more complex organisms. Many species of bacteria play a major role in the cycling of nutrients in ecosystems through aerobic and anaerobic decomposition. Finally, some species form symbiotic relationships with more complex organisms and help these life forms survive in the environment by fixing atmospheric nitrogen.

Bajada: Consecutive series of alluvial fans forming along the edge of a linear mountain range. Surface of this feature undulates in a rolling fashion as one moves from the center of one alluvial fan to another. Normally occurs in arid climates.

Bank-Caving: Collapse of stream bank material into a stream channel.

Barchan Dune: Crescent shaped sand dune that has its long axis transverse to the wind and its crescent tips pointed downwind.

Barometer: Instrument that measures atmospheric pressure.

Barrier Beach: A long and narrow beach of sand and/or gravel that runs parallel to the coastline and is not submerged by the tide.

Barrier Island: Long, narrow islands of sand and/or gravel that are usually aligned parallel to the shore of some coasts.

Basal Sliding: The sliding of a glacier over the surface it rests on. Caused by the gradient of the slope and the weight of the glacier's mass.

Basalt: A dark colored fine grained igneous rock formed from mafic magma.

Basalt Plateau: Extensive continental deposits of basaltic volcanic rock.

Basaltic Magma: Mafic magma that forms basaltic igneous rocks.

Base Flow: Rate of discharge in a stream where only the throughflow and groundwater flow from subsurface aquifers contribute to the overall flow.

Base Level: The subterranean elevation below which a stream cannot vertically erode sediment. For many streams this hypothetical elevation is sea-level.

Basement Rock: Very old granite and metamorphic rocks found in continental crust. These rocks make up the continental shield.

Basic: Substance having a pH greater than 7.

Basin: A topographic rock structure whose shape is concave downwards.

Batholith: A large mass of subsurface intrusive igneous rock that has its origins from mantle magma.

Bay: A body of sheltered water found in a crescent shaped coastal configuration of land.

Bayhead Beach: An extensive deposit of sand and/or gravel in the form of a beach at the back of a bay.

Bay-Mouth Bar: A narrow deposit of sand and/or gravel found across the mouth of a bay.

Beach: The terrestrial interface area in between land and a water body where there are accumulations of unconsolidated sediments like sand and gravel. These deposits are laid down by the action of breaking waves.

Beach Drift: The lateral movement of sediments on a beach when the angles of swash and backwash differ.

Bearing: A system that measures in reference to the cardinal points of a compass in 90 degree quadrants.

Beaufort Wind Scale: Descriptive system that determines wind speed by noting the effect of the wind on the environment. Originally developed for use at sea by Admiral Beaufort of the British Navy in 1806.

Bed: Sedimentary structure that usually represents a layer of deposited sediment.

Bedding Plane: A layer in a series of sedimentary beds that marks a change in the type of deposits.

Bed Load: Portion of the stream load that is carried along the stream bed without being permanently suspend in the flowing water.

Bedrock: Rock at or near (beneath soil and regolith) the Earth's surface that is solid and relatively unweathered.

Benthos: The plant and animal organisms that live on the sea floor. Often divided into two categories: *deep-sea benthos*, below 200 meters and the *littoral benthos*, from 200 meters to the high-water spring tide level.

Bergschrund: A deep crevasse commonly found at the head of an alpine glacier. Forms when the glacial ice pulls away from the mountain side.

Berm: Low hill of sand that forms along coastal beaches.

Bermuda High: High pressure system that develops over the western subtropical North Atlantic. Also called *Azores High*.

Biennial Plant: Plant species that completes its life in two growing seasons.

Bifurcation Ratio: Quantitative ratio determined between the parts of systems that display branching. For example, trees have a main stem that bifurcates into smaller and smaller branches. The ratio between the branches that are derived from a larger branch or main stem is the bifurcation ratio.

Biological Amplification: Increase in concentration of toxic fat-soluble chemicals in organisms at successively higher trophic levels of a grazing food chain or food web because of the consumption of organisms at lower trophic levels.

Biological Weathering: The disintegration of rock and mineral due to the chemical and/or physical agents of an organism.

Biomass: The weight of living tissues usually measured per unit area over a particular time interval. Can include the dead parts of organisms like bark, hair, and nails.

Biome: Largest recognizable assemblage of animals and plants on the Earth. The distribution of the biomes is controlled mainly by climate.

Bioregion: A unique region on the Earth that has distinct soils, landforms, watersheds, climates, native plants, and animals, and/or other particular natural characteristics.

Biosphere: Part of the Earth where life is found. The biosphere consists of all living things, plant and animal. This sphere is characterized by life in profusion, diversity, and clever complexity. Cycling of matter in this biosphere involves not only metabolic reactions in organisms, but also many abiotic chemical reactions. Also called ecosphere.

Biotic Potential: Maximum rate that a population of a given species can increase in size (number of individuals) when there are no limits on growth rate.

Biotite: Rock forming mineral of the mica group.

Bird: Group of warm blooded vertebrate animals whose body is covered with feathers.

Black Body: Is a body that emits electromagnetic radiation, at any temperature, at the maximum possible rate per unit surface area. This body also absorbs all electromagnetic radiation that is intercepted by it.

Blizzard: Winter severe weather condition characterized by strong wind, blowing snow, and cold temperatures.

Blowout Depression: Saucer shaped depressions created by wind erosion. At the leeward end of the feature there usually is a deposit of sand. Blowouts are found in coastal beach areas and in arid and semiarid regions of the world. These features are smaller than a deflation hollow.

Body Wave: Type of seismic wave that travels through the interior of Earth.

Bog: A habitat that consists of waterlogged spongy ground. Common vegetation are sedges and sphagnum moss. Bogs are common in Canada, Russia, and Scandinavia.

Bolson: Is a closed desert basin with no drainage outlet, surrounded by mountains.

Boulder: Large fragment of rock that has a diameter greater than 256 millimeters (200 millimeters in the United Kingdom).

Bora: Term used to describe a katabatic wind in Yugoslavia.

Boreal Forest: High to mid-latitude biome dominated by coniferous forest. Predominnt vegetation of this biome is various species of spruce, fir, pine, and cedars.

Bottomset Bed: Horizontal deltaic deposit of alluvial sediment composed of fine silt and clay.

Bowen Reaction Series: Model that explains the origin of the various types of igneous rocks. It suggests that the presence or absence of particular minerals in igneous rocks depends on the temperature of crystallization and on the magma's original chemical composition.

Brackish: Environment that is influenced by seawater with a salinity less than 35 parts per thousand (usually caused by the presence of an inflow of fresh water).

Braided Stream: Shallow stream channel that is subdivided into a number of continually shifting smaller channels that are separated by bar deposits.

Breaker: The quick collapse of an overextended water wave as it approaches the shoreline. The collapse occurs when the ratio of wave height to wavelength exceeds 1:7. This phenomenon also produces swash.

Breccia: Coarse grained sedimentary rock composed of cemented angular rock fragments.

Brine: Seawater with a salinity greater than 35 parts per thousand. Usually occurs in isolated bodies of seawater that have high amounts of water loss due to evaporation.

British Thermal Unit (Btu): Measurement unit for heat. It is the amount of energy required to raise the temperature of one pound of water one degree from 62 to 63° Fahrenheit. One Btu is equal to 252 and to 1055 joules.

Bromeliad: Plants of the bromeliad family (*Bromeliaceae*). These plants grow from the dry deserts of the subtropics to equatorial tropical rain forests. Many bromeliads grow high up on the branches and trunks of trees in the tropical rainforest. Based on growth habits and other characteristics, Bromeliaceae is divided into the subfamilies *Pitcairnioideae*, *Tillandsioideae*, and *Bromelioideae*.

Brunisol Soil: Soil order (type) of the Canadian System of Soil Classification. This soil type is associated with forest vegetation. It is usually poorly developed and immature. The most identifying trait of this soil is the presence of a brown B horizon.

Calcification: A dry environment soil-forming process that results in the accumulation of calcium carbonate in surface soil layers.

Calcite: Mineral formed from calcium carbonate. Common mineral found in limestone.

Calcium Carbonate: Compound consisting of calcium and

carbonate. Calcium carbonate has the following chemical structure $CaCO_3$.

Caldera: A large circular depression in a volcano.

Caldera Volcano: Explosive type of volcano that leave a large circular depression. Some of these depressions can be as large as 40 kilometers in diameter. These volcanoes form when wet granitic magma quickly rises to the surface of the Earth.

Caliche: An accumulation of calcium carbonate at or near the soil surface.

Calorie: Quantity of energy. Equals the amount of heat required to raise 1 gram of pure water from 14.5 to 15.5° Celsius at standard atmospheric pressure.

Calving: The loss of glacier mass when ice breaks off into a large water body like an ocean or a lake.

Cambrian: Geologic period that occurred from 570 to 505 million years ago. During this period, invertebrates become common in the oceans and the Burgess Shale was formed.

Cambrian Explosion: Great diversification of multicellular life forms in the Earth's oceans that started during the Cambrian about 570 million years ago.

Canadian High: High pressure system that develops in winter over central North America.

Canadian Shield: Very old igneous and metamorphic shield rock that covers much of northern Canada. Created more than two to three billion years ago.

Canadian System of Soil Classification: A hierarchical system that is used in Canada to classify soils. This system has five levels: *order*, *great group*, *subgroup*, *family*, and *series*. At the *order* level, nine types of soils are recognized: brunisol, chernozem, cryosol, gleysol, luvisol, organic, podzol, regosol, and solonetzic.

Canopy Drip: Redirection of a proportion of the rain or snow falling on a plant to the edge of its canopy.

Canyon: Steep-sided valley where depth is considerably greater than width. These features are the result of stream erosion.

Capillary Action: Movement of water along microscopic channels. This movement is the result of two forces: the adhesion and absorption of water to the walls of the channels; and cohesion of water molecules to each other.

Capillary Water: Water that moves horizontally and vertically in soils by the process of capillary action. This water is available for plant use.

Carbohydrate: Is an organic compound composed of carbon, oxygen, and hydrogen atoms. Some examples are sugars, starch, and cellulosc.

Carrying Capacity (K): The maximum size of population of a single species that a certain habitat can support.

Cartography: Field of knowledge that studies map construction. The act of creating a map.

Cascading System: This is a system where we are primarily interested in the flow of energy and/or matter from one element to another and understand the processes that cause this movement. In a cascading system, we do not fully understand quantitative relationships that exist between elements related to the transfer of energy and/or matter.

Catastrophism: General theory that suggests that certain phenomena on the Earth are the result of catastrophic events. For example, the *Biblical Flood* is responsible for sedimentary rock formations and the extinction of the dinosaurs.

Cation: An ion carrying a positive atomic charge.

Cation Exchange: Chemical trading of cations between the soil minerals and organic matter with the soil solution and plant roots.

Cation Exchange Capacity: The capacity of a soil to exchange cations with the soil solution. Often used as a measure of potential soil fertility.

Cave: A natural cavity or recess that is roughly positioned horizontally to the surface of the Earth.

Cellular: Composed of cells. Process occurring between or within cells.

Cellulose: A type of carbohydrate. Primary component used in the construction of plant cell walls.

Celsius Scale: Scale for measuring temperature. In this scale, water boils at 100° and freezes at 0°.

Cenozoic: Geologic era that occurred from 65 million years ago to today.

Central Vent: The main passage way by which volcanic magma travels to the Earth's surface.

Centripetal Force: Force required to keep an object moving in a circular pattern around a center of rotation. This force is directed towards the center of rotation. Common in meteorological phenomena like tornadoes and hurricanes.

C Horizon: Soil horizon normally found below the B horizon and above the R horizon. This layer is composed of weathered bedrock that has not been yet significantly affected by the pedogenic processes.

Chain Reaction (Nuclear): A large number of nuclear fissions, taking place within a certain mass of a fissionable isotope, that release a great quantity of energy in a short time.

Chalk: Form of limestone. This sedimentary rock is composed of the shells and skeletons of marine microorganisms.

Chaparral: A type of plant community common to areas of the world that have a Mediterranean climate (for example, California and Italy). It is characterized by shrubs, shrubby thickets and small trees that are adapted to seasonal dry conditions. Also called *Mediterranean Scrubland.*

Chelate: Organic substances that cause the chemical process of chelation.

Chelation: Chemical weathering process that involves the extraction or metallic cations from rocks and minerals by chelates.

Chemical: One of the millions of different elements and compounds found naturally and synthesized by humans.

Chemical Autotroph: Organism that uses the external energy found in chemical compounds to produce food molecules. The process used to produce food by these organisms is known as chemosynthesis.

Chemical Energy: Energy used or released in chemical reactions.

Chemical Reaction: Reaction between chemicals where there is a change in the chemical composition of the elements or compounds concerned.

Chemical Weathering: Breakdown of rock and minerals into small sized particles through chemical decomposition.

Chinook Wind: The name of a North American wind that occurs on the leeward side of mountains. This wind is warm and has a low humidity.

Chlorofluorocarbons (CFCs): Is an artificially created gas that has become concentrated in the Earth's atmosphere. This very strong greenhouse gas is released from aerosol sprays, refrigerants, and the production of foams. The basic chemical formula for chlorofluorocarbons is $CF_x Cl_x$.

Chlorophyll: Green pigment found in plants and some bacteria used to capture the energy in light through photosynthesis.

Chloroplast: Organelle in a cell that contains chlorophyll and produces organic energy through photosynthesis.

Chromosome: Organic structure that carries an organism's genetic code (DNA).

Cinder Cone Volcano: A small volcano, between 100 and 400 meters tall, made up of exploded rock blasted out of a central vent: at a high velocity. These volcanoes develop from magma of basaltic to intermediate composition.

Circle of Illumination: A line that bisects areas on the Earth receiving sunlight and those areas in darkness. Cuts the spherical Earth into lighted and dark halves.

Circum-Pacific Belt: A zone circling the edge of the *Pacific Ocean* basin where tectonic subduction causes the formation of volcanoes and trenches. Also called the *ring of fire*.

Cirque: Glacially eroded rock basin found on mountains. Most alpine glaciers originate from a cirque.

Cirque Glacier: Small glacier that just occupies a cirque.

Cirrocumulus Clouds: Patchy white high altitude cloud composed of ice crystals. Found in an altitude range from 5,000 to 18,000 meters.

Cirrostratus Clouds: High altitude sheet like clouds composed of ice crystals. These thin clouds often cover he entire sky. Found in an altitude range from 5,000 to 18,000 meters.

Cirrus Clouds: High altitude cloud composed of ice crystals. The appearance of these clouds is white feather like patches, filaments or thin bands. Found in an altitude range from 5,000 to 18,000 meters.

Classification: Process of grouping things into categories.

Clastic Sedimentary Rock: Sedimentary rocks that are formed by the lithification of weathered rock debris that has been physically transported and deposited.

Cleavage: The tendency of some minerals or rocks to break along planes of weakness. This weakness occurs because of the nature of the bonds between mineral grains.

CLIMAP Project: Multiuniversity research project that reconstructed the Earth's climate for the last million years by examining proxy data from ocean sediment cores.

Climate: General pattern of weather conditions for a region over a long period time (at least 30 years).

Climatic Optimum: Warmest period during the Holocene epoch. This period is dated from about 5,000 to 3,000 BC. During this time average global temperatures were 1 to 2° Celsius warmer than they are today.

Climatology: Scientific study of the Earth's climate over long time spans (greater than several days). May also involve the investigation of climate's influence on the biotic and the abiotic environment.

Climax Community: Plant community that no longer undergoes changes in species composition due to succession.

Climograph: Two dimensional graph that plots a location's air temperature and precipitation on times scales that range from a 24 hour period to a year.

Clone: A group of genetically similar plants that have originated by vegetative asexual reproduction from a single parent. Replication of an individual who is genetically identical to its parent.

Closed System: Is is a system that transfers energy, but not matter, across its boundary to the surrounding environment. Our planet is often viewed as a closed system.

Closed Talik: Is a form of localized unfrozen ground (talik) in an area of permafrost. It is completely enclosed by permafrost in all directions.

Cloud: A collection of tiny particles of liquid or solid water occurring above the Earth's surface. Clouds are classified accord to their height of occurrence an shape. The major types of clouds include: Cirrus, Cirrocumulus, Cirrostratus, Altocumulus, Altostratus, Nimbostratus, Stratocumulus, Stratus, Cumulus, and Cumulonimbus.

Colonization: Movement of individuals or propagules of a species to a new territory.

Comet: A large mass of ice and dust that has an orbit around a star.

Commensialism: Biological interaction between tospecies where one species benefits in terms of fitness while they other experiences no effect on its fitness.

Community: Refers to all the populations of interacting species found in a specific area or region at a certain time.

Community Boundary: Spatial edge of a unique community.

Compass: Navigation instrument that uses the Earth's magnetic field to determine direction.

Competitive Exclusion: Situation where no two competitively interacting species can occupy exactly the same fundamental niche indefinitely because of resource limitations. The outcome of this process is the local extinction the species that is a poorer competitor.

Composite Volcano: Volcano created from alternate layers of flows and exploded rock. Their height ranges from 100 to 3,500 meters tall. The chemistry of the magma of these volcanoes is quite variable ranging from basalt to granite.

Composites: Plants of the compositae family (Asteraceae). Common examples of these flowering plants are thistles, dandelion, and sunflowers.

Compound: A compound is the atoms of different elements joined together.

Concrete Space: Actual geographic space in the real world. Geographers approximate this space when they try to represent it in a model or map. This approximation is referred to as abstract space.

Condensation: The change in state of matter from vapor to liquid that occurs with cooling. Usually used in meteorology when discussing the formation of liquid water from vapor. This process releases latent heat energy to the environment.

Condensation Nuclei: Microscopic particle of dust, smoke or salt that allows for condensation of water vapor to water droplets in the atmosphere. Nucleus for the formation of a rain drop. Condensation normally occurs on these particles when relative humidity becomes 100 %. Some condensation nuclei, like salt, are hygroscopic and water can condense on them at relative humidities lower than 100 %.

Conduction: Conduction consists of energy transfer directly from atom to atom and represents the flow of energy along a temperature gradient.

Cone of Depression: Cone shaped depression occurring horizontally across a water table. Causes by excessive removal of groundwater by a surface well.

Cnfined Aquifer: Aquifer between two layers of relatively impermeable earth materials, such as clay or shale.

Confined Groundwater: Groundwater trapped between two impervious layers of rock.

Conglomerate: Coarse grained sedimentary rock composed of rounded rock fragments cemented in a mixture of clay and silt.

Coniferous Vegetation: Cone-bearing vegetation of middle and high latitudes that are mostly evergreen and that have needle-shaped or scale like leaves. Compare with deciduous vegetation.

Conservation Biology: Multidisciplinary science that deals with the conservation of genes, species, communities, and ecosystems that make up Earth's biodiversity. It generally investigates human effects on biodiversity and tries to develop practical approaches to preserving biodiversity and ecological integrity.

Contact Metamorphism: Is the small scale metamorphic alteration of rock due to localized heating. It is usually cause by an igneous intrusion like a sill or a dyke.

Continental Arctic Air Mass (A): Air mass that forms over extensive landmass areas of the high latitudes. In the Northern Hemisphere, these system form only in winter over Greenland, northern Canada, northern Siberia, and the Arctic Basin. Continental Arctic air masses are very cold and extremely dry. These air masses are also very stable.

Continental Slope: Steeply sloping portion of continental crust found between the continental shelf and continental rise.

Continental Tropical Air Mass (cT): Air mass that forms over extensive landmasses areas of the low latitudes. In North America, these system form over southwestern United States and northern Mexico. Continental Tropical air masses are warm and dry in the winter and hot and dry in the summer. These air masses are also generally unstable in the winter and stable in the summer.

Contour (Line): Line on a topographic map that connects all points with the same elevation.

Contour Interval: Difference in elevation between two successive contour lines. The interval at which contours are drawn on a map depends on the amount of the relief depicted and the scale of the map.

Control System: A system that is intelligently controlled by the activities of humans. For example, a dam on a river.

Continuous Permafrost: Form of permafrost that exists across a landscape as an unbroken layer.

Convection: Convection involves the transfer of heat energy by means of vertical mass motions through a medium.

Convection Current: The movement of a gas or a fluid in chaotic vertical mass motions because of heating.

Convectional Lifting: The vertical lifting of parcels of air trough convective heating of the atmosphere. This process can initiate adiabatic processes inside the air parcel.

Convectional Precipitation: Is the formation of precipitation due to surface heating of the air at the ground surface. If enough heating occurs, the mass of air becomes warmer and lighter than the air in the surrounding environment, and just like a hot air balloon it begins to rise, expand and cool. When sufficient cooling has taken place saturation occurs forming precipitation. This process is active in the interior of continents and near the equator forming cumulus clouds and possible later thunderstorms. Rain is usually the precipitation type that is formed, and in most cases this moisture is delivered in large amounts over short periods of time in extremely localized areas.

Convergence: Horizontal inflow of wind into an area. Once at the area, the wind then travels vertically.

Convergence Precipitation: The formation of precipitation due to the convergence of two air masses. In most cases, the two air masses have different climatological characteristics. One is usually warm and moist, while the other is cold and dry. The leading edge of the latter air mass acts as an inclined wall or front causing the moist warm air to be lifted. Of course the lifting causes the warm moist air mass to cool due to expansion resulting in saturation. This precipitation type is common at the mid-latitudes where cyclones form along the polar front. Also called *frontal precipitation*.

Convergent Lifting: The vertical lifting of parcels of air through the convergence of opposing air masses in the atmosphere. This process can initiate adiabatic processes inside the air parcel.

Coordinated Universal Time (UTC): Current official world time reference for civil and scientific purposes. Coordinated

Universal Time is measured from six standard atomic clocks at the *International Bureau of Weights and Measures* (*BIPM*) in Paris, France. Implemented in 1964.

Coral: Simple marine animals that live symbiotically with algae. In the symbiotic relationship, the algae provides the coral with nutrients, while the coral provide the algae with a structure to live in. Coral animals secrete calcium carbonate to produce a hard external skeleton.

Coral Bleaching: Situation where coral lose their colorful symbiotic algae. Thought to be caused by unusually warm water, changes in salinity of ocean seawater, or excessive exposure to ultraviolet radiation.

Coral Reef: Ridge of limestone found generally below the ocean surface. This marine feature is produced by numerous colonies of tiny coral animals, called polyps, that create calcium carbonate structures around themselves for protection. When the corals die, their vacant exterior skeletons form layers that cause the reef to grow. Coral reefs are found in the coastal zones of warm tropical and subtropical oceans.

Core: The core is a layer rich in iron and nickel found in the interior of the Earth. It is composed of two sub-layers: the inner core and outer core. The core is about 7,000 kilometers in diameter.

Coriolis Force: An apparent force due to the Earth's rotation. Causes moving objects to be deflected to the right in the Northern Hemisphere and to the left in the Southern hemisphere. Coriolis force does not exist on the equator. This force is responsible for the direction of flow in meteorological phenomena like mid-latitude cyclones, hurricanes, and anticyclones.

Correlation Coefficient: Statistic that measures the degree of linear association between two variables. Its values vary from between -1 and 1. Perfect positive (the dependent variable increases with an increase in the independent variable) linear association has a correlation coefficient of

1. Perfect negative (the dependent variable decreases with an increase in the independent variable) linear association has a correlation coefficient of -1. Absolutely no association between variables has a value of zero.

Counter-Radiation: Redirection of the Earth's longwave radiation back to the surface because of the greenhouse effect.

Craton: Stable foundation core of the Earth's various plates of continental crust. Composed of the shield and platform.

Crater: Circular depression in the ground surface created by volcanic activity or asteroid impact.

Cretaceous: Geologic period that occurred roughly 65 to 144 million years ago. During this period, the first flowering plant species appear and dinosaurs are at their greatest diversity. Dinosaurs die out at the end of this period.

Crevasse: Opening on a levee that allows for the drainage of water from the floodplain to the stream channel.

Critical Entrainment Velocity: Velocity required to entrain a particular sized particle into the moving medium of air or water.

Crust: Earth's outer most layer of solid rock. Between 7 to 70 kilometers thick. Two types of crust exist: oceanic crust and continental crust.

Cryosol Soil: Soil order (type) of the Canadian System of Soil Classification. This soil is common to high latitude tundra environments. The main identifying feature of this soil is a layer of permafrost within one meter of the soil surface.

Cryostatic Pressure: Pressure exerted on a substance by ice at rest.

Cryotic: Something that is frozen.

Cumulus Cloud: Puffy clouds with relatively flat bases. Cumulus clouds form when moist warm air bubbles vertically escape from the Earth's surface. Found in an altitude range from 300 to 2,000 meters.

Cumulonimbus Cloud: A well developed vertical cloud that often has top shaped like an anvil. These clouds are very dense with condensed and deposited water. Weather associated with this cloud includes: strong winds; hail; lightning; tornadoes; thunder; and heavy rain. When this weather occurs these clouds are then thunderstorms. Can extend in altitude from a few hundred meters above the surface to more than 12,000 meters.

Cuspate Foreland: Is a triangular accumulation of sand and/or gravel located along the oastline. This feature is formed by the joining of two spits.

Cyanobacteria: Bacteria that have the ability to photosynthesize.

Cyclogenesis: Process of cyclone formation, maturation, and death.

Cyclone: Area of low pressure in the atmosphere that displays circular inward movement of air. In the Northern Hemisphere circulation is counterclockwise, while Southern Hemisphere cyclones have clockwise wind patterns.

Cytoplasm: All of the protoplasm in a cell except for what is contained in the nucleus.

Earth Albedo: Is the reflectivity of the Earth's atmosphere and surface combined. Measurements indicate that the average Earth albedo is approximately 30 %.

Earthflow: A rapid type of downslope mass movement that involves soil and other loose sediments. Usually triggered by water saturation from rainfall.

Earthquake: Is a sudden motion or trembling in the Earth. The motion is caused by the quick release of slowly accumulated energy in the form of seismic waves. Most earthquakes are produced along faults, tectonic plate boundaries, or along the mid-oceanic ridges.

Earthquake Focus: Point of stress release in an earthquake.

Earth Revolution: Refers to the orbit of the Earth around the sun. This celestial motion takes 365 1/4 days to complete one cycle. Further, the Earth's orbit around the sun is not circular, but elliptical.

Earth Rotation: Refers to the spinning of the Earth on its polar axis.

Earth Sciences Tradition: Academic tradition in modern Geography that investigates natural phenomena from a spatial perspective.

Easterly Wave: Atmospheric disturbance in the tropical trade winds. Occasionally these systems intensify into hurricanes.

Easting: First measurement of a grid reference used to specific the location of a point on a rectangular coordinate system. The distance measured eastward from the origin of a rectangular coordinate system.

Ebb Tide: Time during the tidal period when the tide is falling. Compare with flood tide.

Eccentricity: Geometric shape of the Earth's orbit. This shape varies from being elliptical to almost circular.

Ecological Niche: Is all of the physical, chemical and biological conditions required by a species for survival, growth and reproduction. Two further abstractions of this concept are the fundamental niche and the ralized niche.

Ecology: The study of the factors that influence the distribution and abundance of species.

Ecosystem: An ecosystem is a system where populations of species group together into communities and interact with each other and the abiotic environment.

Energy: Is defined as the capacity for doing work. Energy can exist the following forms: radiation; kinetic energy; potential energy; chemical energy; atomic energy; electromagnetic energy; electrical energy; and heat energy.

Energy Flux: The rate of energy flow from, into, or through a substance.

Entisols: Soil order (type) of the United States Natural Resources Conservation Service Soil Classification System. Soil of recent development with no or poorly developed soil horizons.

Entrainment: One of three distinct processes involved in erosion. More specifically, it is the process of particle lifting by an agent of erosion.

Entropy: Entropy is the measure of the disorder or randomness of energy and matter in a system.

Environmental Gradient: Spatial gradient where abiotic and biotic factors vary.

Environmental Lapse Rate (ELR): The rate of air temperature increase ordecrease with altitude. The average ELR in the troposphere is an air temperature decrease of 6.5° Celsius per 1000 meters rise in elevation.

Environmental Science: Field of knowledge that studies of how humans and other species interact with one another and with the nonliving environment. It is both a physical and social science that integrates knowledge from a wide range of disciplines, including physics, chemistry, biology, geology, geography, economics, political science, sociology, psychology, and philosophy.

Environmental System: A system where life interacts with the various abiotic components found in the atmosphere, hydrosphere, and lithosphere.

Enzyme: Are types of proteins that are used to facilitate and regulate chemical reactions within cells.

Epiphyte: Type of vegetation that gets its physical support from the branches of other plants. Commonly found in the tropical forests.

Epoch: Geologic time unit that is shorter than a period.

Equator: Location on the Earth that has a latitude of 0°.

Equilibrium: Equilibrium describes the average condition of a system, as measured through one of its elements or attributes, over a specific period of time.

Equinox: Two periods when the declination of the sun is at the equator. The autumnal equinox occurs on September 22 or 23. The vernal equinox occurs on March 21 or 22.

Era: Geologic time unit that is shorter than an eon but longer than a period.

Erg Desert: A region in a desert where sand is very abundant.

Erosion: The removal of weathered sediment or rocks by the forces of wind, water, and ice.

Erosional Landform: Is a landform formed from the removal of weathered and eroded surface materials by wind, water, glaciers, and gravity. This includes landforms with some of the following geomorphic features: river valleys, glacial valleys, and coastal cliffs.

Erratic: A large rock boulder that has been transpoted by glaciers away from its origin and deposited in a region of dissimilar rock.

Esker: Long twisting ridges of sand and gravel found on the Earth's surface. Created when the deposits of subsurface glacial streams are placed on the ground after glacial melting.

Establishment: Subsequent growth and/or reproduction of a colonized species in a new territory.

Estimator: An estimator is any value calculated from the sample data For example, the sample mean is an estimator of the population mean.

Estuary: Somewhat enclosed coastal area at the mouth of a river where nutrient rich fresh water meets with salty ocean water.

Eukaryota: All the organisms with a eukaryote cell type. This group includes animals, plants, fungi, and protists.

Eukaryote: Organisms whose cells have a membrane-bound nucleus and many specialized structures located within their cell boundary. In these organisms, genetic material is organized into chromosomes that reside in the nucleus.

Eustacy: Variations in sea-level that are related to changes in the volume of seawater in the oceans.

Eutrophication: Physical, chemical and biological changes in a water body as a result of the input nitrogen and phosphorus.

Evaporation: Evaporation can be defined as the process by which liquid water is converted into a gaseous state. Evaporation can only occur when water is available. It also requires that the humidity of the atmosphere be less than the evaporating surface (at 100 % relative humidity there is no more evaporation). The evaporation process requires large amounts of energy. For example, the evaporation of one gram of water at a temperature of 100° Celsius requires 540 calories of heat energy (600 calories at 0° Celsius).

Evaporation Fog: A type of fog produced from the advection of cold air over warm water or warm or moist land. This type of fog is sometimes called **steam fog** or **sea smoke**.

Evaporation Pan: Meteorological instrument that is used to measure evaporation rates.

Evaporite: Type of sedimentary rock that is formed from the concentration of dissolved salts through evaporation.

Evapotranspiration: Combined loss of water to the atmosphere via the processes of evaporation and transpiration.

Evolution: Is a process by which species come to possess genetic adaptations to their environment. Its mechanism is natural selection. It also requires genetic mutations.

Exfoliation Dome: A physical weathering feature associated with granite that is the result of the erosion of overburden material and pressure-release. With the release of pressure, layers of rock break off in sheets or shells leaving a dome-like bedrock feature.

Exogenic: Refers to a system that is external to the Earth.

Exosphere: The outermost zone in the Earth's atmosphere. This layer has an altitude greater than 480 kilometers and is primarily composed of hydrogen and helium gas.

Exotic Stream: A stream that has a course that begins in a humid climate and end in an arid climate. Because of reductions in precipitation and and increases in evaporation, the discharge of these streams deceases downslope. Examples of such streams are the Nile and Colorado Rivers.

Experiment: A controlled investigation designed to evaluate the outcomes of causal manipulations on some system of interest.

Exploitation: Form of competition where the indirect effects of the two or more species or individuals reduce the supply of the limiting resource or resources needed for survival.

Explosive Eruption: Volcanic eruption where high-viscosity granite-rich magma causes an explosion of ash and pyroclastic material. This type of eruption is common to composite and caldera volcanoes.

Extrusive Igneous Rock: Igneous rock that forms on the surface of the Earth. Also called *volcanic igneous rock.*

Facilitation: Modification of a system that makes subsequent modifications easier.

Facilitation Model of Succession: This model of succession suggests that the change in plant species dominance over time is caused by modifications in the abiotic environment that are imposed by the developing community. Thus the entry and growth of the later species depends on earlier species preparing the ground.

Fahrenheit Scale: Scale for measuring temperature. In this scale, water boils at 212° and freezes at 32°.

Fall: Season between summer and winter. Astronomically it is the period from the autumnal equinox to the in the Northern Hemisphere.

False Origin: Location of the starting coordinates picked to the south and west of the true origin of a rectangular coordinate system. False origins are used to avoid negative coordinates.

Falsifiation: Falsification is a procedure used in science to test the validity of a hypothesis or theory. It involves stating some output from theory in specific and finding contrary cases among experiments or observations.

Fault: A fracture in rock caused by stress.

Fault Plane: The plane that represents the fracture surface of a fault.

Fault Scarp: The section of the fault plane exposed in a fault. Also called an *escarpment.*

Fission (Nuclear): Process where the mass of an atomic nucleus is made smaller by the removal of subatomic particles. This process releases atomic energy in the form of heat and electromagnetic radiation.

Fissionable Isotope: Isotope that can undergo nuclear fission when hit by a neutron at the right speed. Examples include uranium-235 and plutonium-239.

Flash Flood: A rapid and short-lived increase in the amount of runoff water entering a stream resulting in a flood.

Flocculation: Chemical processes where salt causes the aggregation of minute clay particles into larger masses that are too heavy to remain suspended water.

Flood: Inundation of a land surface that is not normally submerged by water from quick change in the level of a water body like a lake, stream, or ocean.

Flood Tide: Time during the tidal period when the tide is rising. Compare with ebb tide.

Fluid: Substance, gas or liquid, that has the property of flow.

Fluid Drag: Reduction in the flow velocity of a fluid by the frictional effects of a surface.

Fluvial: Involving running water. Usually pertaining to stream processes.

Fold: Wavelike layers in rock strata that are the result of compression.

Folding: The deformation of rock layers because of compressive forces to form folds.

Foliar Leaching: Process in which water from precipitation removes plant nutrients from the surface of leaves.

Foliation: Process where once randomly distributed platy minerals in a rock become reoriented, because of metamorphism, in a parallel manner.

Food Chain: Movement of energy through the trophic levels of organisms. In most ecosystems, this process begins with photosynthetic autotrophs (plants) and ends with carnivores and detritivores.

Foot Wall: The bottommost surface of an inclined fault.

Force: Process that changes the state of rest or motion of a body.

Force of Acceleration: Force resulting in the speed of a moving body to increase.

Forest: Ecosystem dominated by trees. Major forest biomes include tropical evergreen forest, tropical savanna, deciduous forest, and boreal forest.

Foreset Bed: Deltaic deposit of alluvial sediment that is angled 5 to 25° from horizontal. Most of the delta is made up of these deposits.

Foreshock: Small earth tremors that occur seconds to weeks before a significant earthquake event.

Fossil: Geologically preserved remains of an organism that lived in the past.

Freeze-Thaw Action: Processes associated with daily and seasonal cycles of freezing and melting.

Fresh Water: Water that is relatively free of salts.

Friction: Resistance between the contact surfaces of two bodies in motion.

Frictional Force: Force acting on wind near the Earth's surface due to frictional roughness. Causes the deceleration of wind.

Front: Transition zone between air masses with different weather characteristics.

Frost Wedging: A process of physical weathering in which water freezes in a crack and exerts force on the rock causing further rupture.

Fujita Tornado Intensity Scale: Tornado classification system developed by T. Theodore Fujita. This system six levels from F0 to F5. These levels are based on the estimated speed of the tornado's winds from proxy information like property damage.

Fusion (Nuclear): Process where the mass of an atomic nucleus is made larger by the addition of subatomic particles. This process releases atomic energy in the form of heat and electromagnetic radiation.

Gabbro: An intrusive igneous rock that develops from mafic magma and whose mineral crystals are coarse. Mineralogically this rock is identical to basalt.

Galaxy: An assemblage of millions to hundreds of billions of stars.

Gamma Radiation: A type of ionizing, electromagnetic radiation that readily penetrates the body tissues of organisms. Has a wavelength less than 0.03 nanometers.

Gap: A spatial opening in a plant community. Can be caused by natural death or by some other abiotic or biotic disturbance.

Gas: A state of matter where molecules are free to move in any direction they like. The state of matter where the substance completely fills any container that it occupies.

Gelisols: Soil order (type) of the United States Natural Resources Conservation Service Soil Classification System. This soil is common to high latitude tundra environments. The main identifying feature of this soil is a layer of permafrost within one meter of the soil surface.

Gene(s): Organic material that allows organisms to pass on the inheritance of adaptations or traits. In most organisms these adaptations are coded through the organic molecule DNA. New adaptations appear by way of mutations.

Gene Frequency: Frequency of alleles at an individual or population level.

Gene Pool: Sum total of all the genes found in the individuals of the population of a particular species.

Geodesy: The science that measures the surface features of the Earth.

Geographical Coordinate System: System that uses the measures of latitude and longitude to locate points on the spherical surface of the Earth.

Geographic Range: Spatial distribution of a species. The geographic ranges of species often fluctuate over time.

Geography: The study natural and human constructed phenomena relative to a spatial dimension.

Geoid: True shape of the Earth, which deviates from a perfect sphere because of a slight bulge at the equator.

Geology: The field of knowledge that studies the origin, structure, chemical composition, and history of the Earth and other planets.

Geomorphic Thrshold: The amount of slow accumulated change a landform can take before it suddenly moves into an accelerated rate of change that takes it to a new system state.

Geomorphology: The field of knowledge that investigates the origin of landforms on the Earth and other planets.

Geostationary Orbit: Satellite that has an orbit that keeps it over the same point on the Earth at all times. This is accomplished by having the satellite travel in space at the same angular velocity as the Earth.

Geostrophic Wind: Horizontal wind in the upper atmosphere that moves parallel to isobars. Results from a balance between pressure gradient force and Coriolis force.

Germination: The beginning of vegetative growth of a plant from a seed.

Glacial Drift: A generic term applied to all glacial and glaciofluvial deposits.

Glacial Ice: A very dense form frozen water that is much harder than snow, névé, or firn.

Glacial Lake: A natural impoundment of meltwater at the front of a glacier.

Glacial Milk: Term used to describe glacial meltwater which has a light colored or cloudy appearance because of clay-sized sediment held in suspension.

Glacial Surge: A rapid forward movement of the snout of a glacier.

Glacial Trough: A deep U-shaped valley with steep valley walls that was formed from glacial erosion. At the base of many of these valleys are cirques.

Glacial Uplift: Upward movement of the Earth's crust following isostatic depression from the weight of the continental glaciers.

Glacial Valley: Valley that was influenced by the presence of glaciers. The cross-section of such valleys tends to be U-shaped because of glacial erosion. Similar to glacial trough.

Glacier: A large long lasting accumulation of snow and ice that develops on land. Most glaciers flow along topographic gradients because of their weight and gravity.

Glaciofluvial: Geomorphic feature whose origin is related to the processes associated with glacial meltwater.

Glaze: Coating of ice that forms when rain falls on a surface with a temperature below freezing.

Global Warming: Warming of the Earth's average global temperature because of an increase in the concentration of greenhouse gases. A greater concentration in greenhouse gases in the atmosphere is believed to result in an enhancement of the greenhouse effect.

Glucose: Simple six-carbon sugar. The chemical formula for glucose is $C_6H_{12}O_6$.

Gneiss: A metamorphosed coarse grained igneous rock. In this rock you get the recrystallization of quartz, feldspars, micas and amphiboles into bands.

Gradient: The steepness of a slope as measured in degrees, percentage, or as a distance ratio (*rise/run*).

Gradient Wind: Horizontal wind in the upper atmosphere that moves parallel to curved isobars. Results from a balance between pressure gradient force, Coriolis force, and centripetal force.

Granite: Medium to coarse grained igneous rock that is rich in quartz and potassium feldspar. Derived from felsic magma.

Grass: Type of plant that has long slender leaves that extend from a short stem or the soil surface.

Grassland: Ecosystem whose dominant species are various types of grass. Found in regions where average precipitation is not great enough to support the growth of shrublands or forest.

Graupel: A type of precipitation that consists of a snow crystal and a raindrop frozen together. Also called snow pellets.

Gravel: A term used to describe unconsolidated sediments composed of rock fragments. These rock fragments have a size that is greater than 2 millimeters.

Gravitational Water: Water that moves through soil due to gravitational forces. Soil water in excess of hygroscopic water and capillary water.

Gravity: Is the process where any body of mass found in the universe attracts other bodies with a force proportional to the product of their masses and inversely proportional to the distance that separates them. First proposed by Sir Issac Newton in 1686.

Grazing Food Chain: Model describing the trophic flow of organic energy in a community or ecosystem.

Great Circle: An imaginary circle drawn on the Earth's surface that has its center synchronize to the center of the planet. The equator is a great circle.

Hadean: Geologic eon that occurred from 3800 to 4600 million years ago. The Earth's oldest rocks date to the end of this time period.

Hadley Cell: Three-dimensional atmospheric circulation cell located at roughly 0 to 30° North and South of the equator. The Hadley cell consists of rising air (intertropical convergence zone) at the equator and descending air (subtropical highs) at 30° North and South.

Hanging Wall: The topmost surface of an inclined fault.

Hardpan: Impervious layer found within the soil. It can result from the precipitation of iron, illuviation of clay or the cementing of sand and gravel by calcium carbonate precipitates.

Hazard: Phenomenon which can cause loss of life, injury, disease, economic loss, or environmental damage.

Headlands: A strip of land that juts seaward from the coastline. This feature normally bordered by a cliff.

Headwaters: Upper portion of stream's drainage system.

Heat: Heat is defined as energy in the process of being transferred from one object to another because of the temperature difference between them. In the atmosphere, heat is commonly transferred by conduction, convection, advection, and radiation.

Heat Capacity: Is the ratio of the amount of heat energy absorbed by a substance compared to its corresponding temperature rise.

Heat Energy: A form of energy representing aggregated internal energy of motions of atoms and molecules in a body.

Heat Island: The dome of relatively warm air which develops over the center of urbanized areas.

Helical Flow: Movement of water within a stream that occurs as spiral flows.

Herb: A nonwoody angiosperm whose above ground vegetation dies off seasonally.

Heredity: The transmission of behavioral, physiological and morphological characteristics from parent to offspring.

High Pressure: An area of atmospheric pressure within the Earth's atmosphere that is above average. If this system is on the Earth's surface and contains circular wind flow and enclosed isobars it is called an anticyclone.

Histosols: Soil order (type) of the United States Natural Resources Conservation Service Soil Classification System. This soil is composed more than 30 % organic matter as a result of saturated environmental conditions.

Holistic: Concerned with complete system.

Holocene Epoch: Period of time from about 10,000 years ago to today. During this period glaciers retreated because of a warmer global climate. Time of modern humans.

Horn: Pyramidal peak that forms when several cirques erode a mountain from three or more sides.

Horst Fault: A fault that is produced when two reverse faults cause a block of rock to be push up.

Hot Spot: A volcanic area on the surface of the Earth created by a rising plume of magma.

Host: Organism that develops disease from a pathogen or is being feed on by a parasite.

Human Geography: Field of knowledge that studies human-made features and phenomena on the Earth from a spatial perspective. Subdiscipline of Geography.

Human-Land Tradition: Academic tradition in modern Geography that investigates human interactions with the environment.

Humidity: A general term used to describe the amount of water vapor found in the atmosphere.

Humus: Dark colored semi-soluble organic substance formed from decomposition of soil organic matter.

Hurricane: An intense cyclonic storm consisting of an organized mass of thunderstorms that develops over the warm oceans of the tropics. To be classified as a hurricane, winds speeds in the storm must be greater than 118 kilometers per hour.

Hydration: A form of chemical weathering that involves the rigid attachment of H+ and OH- ions to the atoms and molecules of a mineral.

Hydraulic Gradient: The slope of the water table or aquifer. The hydraulic gradient influences the direction and rate of groundwater flow.

Hydrocarbon: Organic compound composed primarily of hydrogen and carbon atoms. An example of a hydrocarbon is methane (CH_4).

Hydrograph: A graph describing stream discharge over time.

Hydrologic Cycle: Model that describes the movement of water between the hydrosphere, lithosphere, atmosphere, and biosphere.

Hydrostatic Pressure: Force caused by water under pressure.

Hygrometer: An instrument for measuring atmospheric humidity.

Hygroscopic: Substances that have the ability to absorb water and therefore accelerate the condensation process.

Hygroscopic Coefficient: Maximum limit of hygroscopic water around the surface of a soil particle.

Hygroscopic Water: Water held within 0.0002 millimeters of the surface of a soil particle. This water is essentially non-mobile and can only be removed fro the soil through heating.

Hyphae: Thread like structures found on a fungus.

Hypothesis: A tentative assumption that is made for the purpose of empirical scientific testing. A hypothesis becomes a theory when repeated testing and evidence suggests the hypothesis has a strong chance of being correct.

Ice Cap: Large dome-shaped glacier found covering a large expanse of land. Smaller than an ice sheet.

Ice Field: Large level area of glacial ice found covering a large expanse of land. Similar in size to an ice cap but does not have a dome-shape.

Ice Fog: A fog that is composed of small suspended ice crystals. Common in Arctic locations when temperatures are below -30° Celsius and a abundant supply of water vapor exists.

Ice Jam: The accumulation of ice at a specific location along a stream channel. Can cause the reduction of stream flow down stream of the obstruction and flooding upstream.

Icelandic Low: Subpolar low pressure system found near Iceland. Most developed during the winter season. This large-scale pressure system spawns mid-latitude cyclones.

Ice Lense: Horizontal accumulation of permanently frozen ground ice.

Ice Pellets: A type of precipitation. Ice pellets or sleet are transparent or translucent spheres of frozen water that fall from clouds. Ice pellets have a diameter less than 5 millimeters. To form, this type of precipitation requires an environment where raindrops develop in an atmosphere where air temperature is above freezing. These raindrops then fall into a lower layer of air with freezing temperatures. In this lower layer of cold air, the raindrops freeze into small ice pellets. Like freezing rain, an air temperature inversion is required for development of ice pellets.

Ice Sheet: A dome-shaped glacier covering an area greater than 50,000 square kilometers. Greenland and Antarctica are considered ice sheets. During the glacial advances of the Pleistocene ice sheets covered large areas of North America, Europe, and Asia. Larger than an ice cap.

Illuviation: Deposition of humus, chemical substances, and fine mineral particles in the lower layers of a soil from upper layers because of the downward movement of water through the soil profile. Compare with eluviation.

Immigration: Migration of an organism into an area for the purpose of changing its residence permanently. Compare with emigration.

Inceptisols: Soil order (type) of the United States Natural Resources Conservation Service Soil Classification System. These soils have horizons in the early stages of pedogenesis. Many of these soils are found on glacial drift deposits.

Independent Variable: Variable in a statistical test that is thought to be controlling through cause and effect the value of observations in another dependent variable modeled in the test.

Index Contour: Contour line that is accentuated in thickness and is often labeled with the appropriate measure of elevation. Index contours occur every four or fifth contour interval and help the map user read elevations on a map.

Indicator Species: Species that can be used as a early indicator of environmental degradation to a community or an ecosystem. Compare with immigrant species, keystone species, and native species.

Induction: Inference of a generalized conclusion from particular instances. In a science like Physical Geography, inductive reasoning would involve the development of a theory to explain previously collected facts or observed phenomenon.

Industrial Revolution: Major change in the economy and society of humans brought on by the use of machines and the efficient production of goods. This period in human history began in England in the late 18th centur.

Inferential Statistics: Statistical test that makes generalizations about a population based of the numeric information obtained from a sample based on the laws of probability.

Infiltration Capacity: The ability of a soil to absorb surface water.

Inorganic: Non-living thing. Usually refers to the physical and chemical components of an organism's environment. Some times called abiotic.

Inosilicate: Subclass of the silicate class of minerals. Inosilicates have two distinct forms: single and double chain silicates.

Inselberg: A German term used to describe a steep-sided hill composed of rock that rises from a pediplain.

Insolation: Direct or diffused shortwave solar radiation that is received in the Earth's atmosphere or at its surface.

Insolation Weathering: Form of physical weathering. Involves the physical breakdown of minerals and rock due to thermal expansion and contraction.

Interaction (biological): Mutual or reciprocal influence between two or more similar organisms or individuals of different species. Major biotic interactions are: competition, mutualism, predation, parasitism, amensalism, and commensialism.

Interference: Form of competition where an individual directly prevents the physical establishment of another individual in a portion of a habitat.

Interglacial: Period of time during an ice age when glaciers retreated because of milder temperatures.

Intermittent Stream: A stream that flows only for short periods over a year. Flow events are usually initiated by rainfall.

International Date Line: A line drawn almost parallel to the 180 degree longitude meridian that marks the location where each day officially begins. The location of the International Date Line was decided upon by international agreement.

Interspecific Interaction: An interaction between different species.

Intertropical Convergence Zone (ITCZ): Zone of low atmospheric pressure and ascending air located at or near the equator. Rising air currents are due to global wind convergence and convection from thermal heating. Location of the thermal equator.

Intraspecific Interaction: An interaction occurring between the individuals that make up a single species.

Inversely Proportional: Cause and effect relationship between two variables where a positive or negative change in the quantity of one causes a predictable opposite change in quantity in the other.

Ion: An atom, molecule or compound that carries either a positive (cation) or negative (anion) electrical charge.

Ionizing Radiation: The emission of alpha or beta particles or gamma rays from radioisotopes. These emitted particles can dislodge one or more electrons from atoms they strike. The free electrons can form charged ions in living tissue that can react with and damage cells.

Isobar: Lines on a map joining points of equal atmospheric pressure.

Isolated System: A system that has no interactions beyond its boundary layer. Many controlled laboratory experiments are this type of system.

Isoline: Lines on a map joining points of equal value.

Isostatic Rebound: The upward movement of the Earth's crust following isostatic depression.

Isotherm: Lines on a map joining points of equal temperature.

Isotope: Form of an element where the number of neutrons in its atomic nucleus is different than the number of protons.

Isotopic Dating: Dating technique used to determine the age of rock and mineral through the decay of radioactive elements.

Joint: A fracture in a rock where no movement has taken place or where no movement has taken place perpendicular to the surface of the fracture. Imporant in rock weathering because it increases the exposed surface area.

Joule: Unit for measuring energy. One joule is the energy used by a force of one Newton in moving its point of application in the direction of the force *one meter*.

Katabatic Wind: Any wind blowing down the slope of a mountain.

Kelvin Scale: Scale for measuring temperature. In this scale, absolute zero is 0 Kelvins, water boils at 373.15 Kelvins and freezes at 273.15 Kelvins.

Kettle Hole: Depression found in glacial deposits. Created when a piece of ice from a retreating glacier becomes embedded in soft glacial till or glacial drift deposits. Many are filled with water to form a small lake or pond.

Kettle Moraine: An area of glaciofluvial influenced moraine deposits pitted with kames and kettle holes.

Keystone Species: Species that interacts with a large number of other species in a community. Because of the interactions, the removal of this species can cause widespread changes to community structure. Compare with immigrant species, indicator species, and native species.

Kinetic Energy: The energy due to motion.

Kingdom: Top most level of the common system used to classify life. Generally, five kingdoms are recognized: Monera, Protista, Fungi, Animalia, and Plantae.

Landslide: Term used to describe the downslope movement of soil, rock, and other weathered materials because of gravity.

Landward: Positioned or located away from a water body but towards the land.

Langley: Unit of the intensity of radiation measured per minute and equal to one calorie.

La Nina: Condition opposite of an El Nino. In a La Nina, the tropical Pacific trade winds become very strong and an abnormal accumulation of cold water occurs in the central and eastern Pacific Ocean.

Latent Heat of Condensation: The amount of heat energy release to the environment when a gas changes its state to a liquid. For one gram of water, the amount of heat energy released is 540 calories at a temperature of 100° Celsius.

Latent Heat of Vaporization: The amount of heat energy

required from the environment to change the state of a liquid to a . For one gram of water, the amount of heat energy required is 540 calories at a temperature of 100° Celsius.

Lateral Moraine: Moraine that is found along the sides of a glacier. Commonly found on glaciers that occupy a valley.

Laterite: Hard subsurface deposit of oxides of aluminum and iron found in tropical soils where the water table fluctuates with seasonal changes in precipitation.

Laurasia: Northern section of Pangaea.

Lava: Molten magma released from a volcanic vent or fissure.

Lava Flow: Stream of lava flowing from a volcanic vent.

Law of Basin Areas: Morphometric relationship observed in the mean basn area size of stream segments of a particular classification order in stream channel branching. Proposed by R.E. Horton.

Law of Conservation of Energy: This law states that energy can be transferred from one system to another in many forms, however, it can not be created nor destroyed. Thus, the total amount of energy available in the universe is constant.

Law of Stream Lengths: Morphometric relationship observed in the cumulative size of stream segment lengths in stream channel branching. Proposed by R.E. Horton.

Law of Stream Number: Morphometric relationship observed in the number of stream segments of a particular classification order in stream channel branching. Proposed by R.E. Horton.

Law of the Minimum: This biological law suggests that organisms are normally limited by only one single physical factor that is in shortest supply relative to demand.

Leachate: Solution containing material leached from a soil.

Leaching: Process in which water removes and transports soil humus and inorganic nutrients in solution.

Leaf Drip: The rain water that fall to the ground surface from plant leaves after it has been intercepted by these structures.

Lee: Side of a slope that is opposite to the direction of flow of ice, wind, or water. Opposite of stoss.

Leeward: Downwind side of an elevated area like a mountain. Opposite of windward.

Legume: Angiosperm plant species that is a member of the Fabaceae (Pea or Bean) family. These plants form symbiotic relationships with specific bacteria species for the purpose of acquiring nitrogen for growth.

Levee: Ridge of coarse deposits found alongside the stream channels and elevated above the floodplain. Forms from the deposition of sediment during floods.

Liana: Species of plant that uses the support of wood plants to elevate its leaves above the forest canopy.

Lichen: Organism that consists of a symbiotic joining of a species of fungi and a species of algae.

Light: A humanly visible form of electromagnetic radiation. This radiation has a wavelength between 0.40 and 0.71 micrometers (μm).

Lightning: Visible discharge of electricity created by thunderstorms.

Limiting Factor: Abiotic condition that most controls the growth of a species. For most terrestrial plants this condition is the supply of the nutrient nitrogen in the soil.

Lipid: Is an organic compound composed of carbon atoms that have two hydrogen atoms attached. Lipids are commonly known as fats and oils, and belong to the family of molecules known as hydrocarbons.

Liquid: A state of matter where molecules have the ability to flow and the surface of this mass displays the property of surface tension.

Lithification: Process by which sediments are consolidated into sedimentary rock.

Lithosphere: Is the solid inorganic portion of the Earth (composed of rocks, minerals, and elements). It can be regarded as the outer surface and interior of the solid Earth.

Litter: Accumulation of leaves, twigs and other forms of organic matter on the soil surface. In most soils, the surface layer of litter is at various stages of decomposition.

Litterfall: Movement of leaves, twigs and other forms of organic matter from the biosphere to the litter layer found in soil.

Little Climatic Optimum: Time period from 900 - 1200 AD. Warmest period since the Climatic Optimum.

Little Ice Age: Time period from 1550 to 1850 AD. During this period, global temperatures were at their coldest since the beginning of the Holocene.

Loam: A soil that contains a roughly equal mixture of clay, sand, and silt. Good for growing most crops.

Lobe: A tongue-like extension of some material. For example, the ice lobe of an alpine glacier.

Location: A term used in geography that deals with the relative and absolution spatial position of natural and human-made phenomena.

Loess: Deposits of silt laid down by aeolian processes over extensive areas of the mid-latitudes during glacial and postglacial times.

Logarithmic Scale: Measurement scale based on logarithms. Values increase on this scale exponentially.

Longshore Drift: The movement and deposition of coastal sediments because of longshore currents.

Longshore Transport: The transport of sediment in water parallel to a shoreline.

Long Wave: A large wave in the polar jet stream and the westerlies that extends from the middle to the upper troposphere. Often associated with the formation of a mid-latitude cyclone at the ground surface. Contrasts with short waves.

Macronutrient: Nutritional element required by an organism in relatively large quantities.

Mafic Magma: Magma that is relative poor in silica but rich in calcium, magnesium, and iron content. This type of magma solidifies to form rocks relatively rich in calcium, magnesium, and iron but poor in silica.

Magma: Molten rock originating from the Earth's interior.

Magma Plume: A rising vertical mass of magma originating from the mantle.

Magnetic Declination: The horizontal angle between true north and magnetic north or true south and magnetic south.

Magnetic Field: The space influence by magnetic force. The Earth's magnetic field is believed to be generated by the planet's core.

Magnetic Reversal: A change in the polarity of the Earth's magnetic field. In the past 4 million years there have been nine reversals.

Magnetosphere: Zone that surrounds the Earth that is influenced by the Earth's magnetic field.

Mantle: Layer of the Earth's interior composed of mostly solid rock that extends from the base of crust to a depth of about 2,900 kilometers.

Map: An abstraction of the real world that is used to depict, analyze, store, and communicate spatially organized information about physical and cultural phenomena.

Map Projection: Cartographic process used to represent the Earth's three-dimensional surface onto a two-dimension map. This process creates some type of distortion artifact on the map.

Mass: Refers to the amount of material found in an object (usually of unit volume).

Mass Balance: The relative balance between the input and output of material within a system.

Mass Extinction: A catastrophic, widespread perturbation where major groups of species become extinct in a relatively short time compared to normal background extinctions.

Mass Movement: General term that describes the downslope movement of sediment, soil, and rock material.

Mass Wasting: General term that describes the downslope movement of sediment, soil, and rock material.

Mean: Statistical measure of central tendency in a set of data. The mean is calculated by adding all of the data values and dividing this quantity by the total number of data values.

Mean Sea-Level: The average height of the ocean surface as determined from the mean of all tidal levels recorded at hourly intervals.

Mean Solar Day: Time it takes to complete one Earth rotation relative to the position of the sun (for example, from midnight to midnight). This measurement takes 24 hours and is longer than a sidereal day because t includes the effect of the Earth's movement (Earth revolution) around the sun.

Meltwater: Water produced from the melting of snow and/or glacial ice.

Meridional Transport: Transport of atmospheric and oceanic energy from the equator to the poles.

Mesa: A flat topped hill that rises sharply above the surrounding landscape. The top of this hill is usually capped by a rock formation that is more resistant to weathering and erosion.

Mesocyclone: A cylinder of cyclonically flowing air that form vertically in a severe thunderstorm. They measure about 3 to 10 kilometers across. About 50 % of them spawn tornadoes.

Mesopause: Thin boundary layer found between the mesosphere and the thermosphere. It is usually found at an average altitude of 80 kilometers. Coldest temperatures in the atmosphere are found at the mesopause.

Mesophyte: Plants that have moderate water requirements.

Metabolism: Describes all of the enzymatic reactions performed by the cells of an organism.

Metamorphic Rock: A rock that forms from the recrystallization of igneous, sedimentary or other metamorphic rocks through pressure increase, temperature rise, or chemical alteration.

Metamorphism: Process that creates metamorphic rocks.

Metasomatic Metamorphism: Form of metamorphism that causes the chemical replacement of elements in rock minerals when gases and liquids permeate into bedrock.

Meteorology: The scientific study of the atmosphere and its associated phenomena.

Mid-Oceanic Ridge: Chain of submarine mountains where oceanic crust is created from rising magma plumes and volcanic activity. Also associated with this feature is plate divergence which creates a rift zone.

Migration: Movement of organisms in an intentional way between two points in space. Many migrations are seasonal.

Milankovitch Theory: Theory proposed by Milutin Milankovitch that suggests that changes in the Earth's climate are cause by variations in solar radiation received at the Earth's surface. These variations are due to cyclical changes in the geometric relationship between the Earth and the sun.

Milky Way Galaxy: Aggregation of about 400 billion stars in a flattened, disk-shaped structure in space. Our solar system is found in this structure.

Mississippian: Geologic period that occurred roughly 320 to 360 million years ago. During this period, insects undergo major speciation and ferns first appear. Trees become a dominant plant form on continents.

Native Species: Species that normally exists and reproduces in a specific region of the Earth. Compare with immigrant species, indicator species, and keystone species.

Natural Gas: Hydrocarbon based gas, mainly composed of methane, commonly found in the pores of sedimentary rocks of marine origin.

Natural Hazards: Natural phenomena that produce negative effects on life. The study of the hazards of natural phenomena.

Natural Ionizing Radiation: Ionizing radiation that comes from natural sources in the environment.

Natural Selection: Environment's influence on the reproductive success of individuals in a population. It results in the exclusion of maladapted genetic traits found within individuals in a population.

Neap Tide: Tide that occurs every 14 to 15 days and coincides with the first and last quarter of the moon. This tide has a small tidal range because the gravitational forces of the moon and sun are perpendicular to each other. Contrasts with spring tide.

Needle Ice: A form of periglacial ground ice that consists of groups ice slivers at or immediately below the ground surface. Needle ice is about a few centimeters long.

Negative Feedback: Change in the state of a system that counteracts the measured effect of the initial alteration.

Oblique Aerial Photograph: Photograph taken from a non-perpendicular angle from a platform in the atmosphere.

Obliquity: Tilt of the Earth's polar axis as measured from the perpendicular to the plane of the Earth's orbit around the sun. The angle of this tilt varies from 22.5 to 24.5° over a 41,000 year period. Current obliquity is 23.5°.

Obsidian: Glassy dark colored volcanic rock. Usually composed of rhyolite.

Occluded Front: A transition zone in the atmosphere where an advancing cold air mass sandwiches a warm air mass between another cold air mass pushing the warm air into the upper atmosphere.

Ocean: A body of saline water found occupying all or part of the Earth's ocean basins. There are five recognized oceans: *the Atlantic*, *the Southern Ocean*, *the Pacific Ocean*, *the Indian Ocean*, and *the Arctic Ocean*.

Ocean Basin: Part of the Earth's outer surface that is comprised of the ocean floor, mid-oceanic ridges, continental rise, and continental slope. The ocean basins are filled with saline water that makes up the oceans.

Ocean Current: Large scale horizontal flow of ocean water that is persistent and driven by atmospheric circulation.

Ocean Floor: Flat plain found at the bottom of the ocean. The ocean floor represents the surface of the oceanic crust. The ocean floor lies between the mid-oceanic ridges and the trenches, usually 5,000 to 7,000 meters below the ocean surface. Also called the abyssal plain.

Oceanic Plate: A rigid, independent segment of the lithosphere composed of mainly basalt that floats on the viscous plastic asthenosphere and moves over the surface of the Earth. The Earth's oceanic plates are an average 75 kilometers thick and were formed less than several hundred million years ago at one of the Earth's mid-oceanic ridges.

Oceanography: The scientific study of phenomena found in the world's oceans.

Ocean Trench: Deep depression found at the edge of the ocean floor. Represents area of tectonic plate subduction.

O Horizon: Topmost layer of most soils. It is composed mainly of plant litter and humus.

Oil: Hydrocarbon based liquid commonly found in the pores of sedimentary rocks of marine origin.

Paternoster Lakes: A linear series of mountain valley lakes that are formed from glacial erosion. They form behind glacial moraines or in glacially carved out rock basins. The name of this feature is related to the series of lakes looking like a string of beads.

Pathogen: Microscopic parasite organism that causes disease in a host. Disease causes the host to be less fit and may eventually cause premature death.

Patterned Ground: Term used to describe a number of surface features found in periglacial environments. These features can resemble circles, polygons, nets, steps, and stripes. The development of some of these shapes is thought to be the result of freeze-thaw action.

Peak Annual Flow: The largest discharge produced by a stream during a one year period.

Peat: Partially decomposed remains of plants that once flourished in a waterlogged environment.

Pebbles: A rounded piece of rock that is larger than gravel.

Pediment: A gradually sloping bedrock surface located at the bae of fluvial-eroded mountain range. Found in arid locations and normally covered by fluvial deposits.

Pediplain: An arid landscape of little relief that is occasionally interrupted by the presence of scattered inselbergs. Formed by the coalescence of several pediments.

Pedogenic Regime: The particular soil forming process that operates in a certain climate. Some of the main processes are: laterization, salinization, podzolization, calcification, and gleization.

Pedon: A basic soil sampling unit. Often viewed as a soil profile that goes from the surface to a depth where soil meets bedrock.

Pennsylvanian: Geologic period that occurred roughly 286 to 320 million years ago. During this period, the first reptiles and winged insects appear.

Perched Water Table: Water table that is positioned above the normal water table for an area because of the presence of a impermeable rock layer.

Percolation: Vertical movement or infiltration of water from the Earth's surface to its subsurface. Movement usually stops when the flowing water reaches the water table.

Perennial Plant: Plant species that lives for more than two years.

Peridotite: Coarse grained ultramafic igneous rock composed mainly of olivine and pyroxene. The mantle is though to be composed primarily of this rock type.

Periglacial: Landforms created by processes associated with intense freeze-thaw action in an area high latitude areas or near an alpine or continental glacier.

Pioneer Community: Community dominated by pioneer species of plants.

Pioneer Species: Plant species that dominate a community in the early stages of succession.

Pitted Topography: Landscape characterized by numerous kettle holes on a glacial outwash plain.

Place: A term used in geography that describes the factors that make the location of natural and human-made phenomena unique.

Pollution: Physical, chemical, or biological change in the characteristics of some component of the atmosphere, hydrosphere, lithosphere, or biosphere that adversely influences the health, survival, or activities of humans or other living organisms.

Quantitative Revolution: Time in the history (after 1950) of physical geography when measurement became the central focus of research. This measurement was used primarily for hypothesis testing. With measurement came mapping, models, statistics, and mathematics. Researchers began investigating process rather than mere description of the environment.

Quartz: Mineral with the chemical formula SiO_2. Quartz is common in continental crust but rare in oceanic crust.

Quartzite: Metamorphic rock rich in quartz created by the recrystallization of sandstone.

Quaternary: Geologic period that occurred roughly 1.6 million years ago to today. During much of this period continental glaciers in the Northern Hemisphere covred large regions of land surface in the high and mid-latitudes. *Homo sapiens* appear about 200,000 years BP (before present) and become the first species to alter the Earth's environment on a large-scale.

Radiant Energy: Energy in the form of electromagnetic waves. In some cases it refers to the radiation emitted from the sun.

Radiation: The emission and propagation of energy in the form of electromagnetic waves.

Radiation Fog: A type of fog that is also called ground fog. Radiation fog is generated by near surface cooling by radiation loss during the evening hours. For the fog to develop, the overnight cooling must cause saturation occur. This type of fog is normally quite shallow.

Radioactive Decay: Natural decay of the nucleus of an atom where alpha or beta particle and/or gamma rays are released at a fixed rate.

Radioisotope or **Radioactive Isotope**: A unstable isotope of an element. This material decays spontaneously and releases subatomic particles and electromagnetic energy.

Radiometer: General name for an instrument used to measure radiation over a specific wavelength range.

Rain: A form of precipitation. It is any liquid deposit that falls from clouds in the atmosphere to the ground surface. Rain normally has a diameter between than 0.5 and 5.0 millimeters.

Raindrop Impact: Force exerted by a falling raindrop on a rock, sediment, or soil surface.

Rain Gauge: Instrument that measures the rain that falls at a location over a period of time.

Rainshadow Effect: Reduction of precipitation commonly found on the leeward side of a mountain. The reduction in precipitation is the result of compression warming of descending air.

Rainsplash: Soil erosion caused from the impact of raindrops.

Rainwash: The erosion of soil by overland flow. Normally occurs in concert with rainsplash.

Random: Process or event that occurs by chance.

Range: A statistical measure of the dispersion of observation values in a data set. Determined by taking the difference between the largest and the smallest observed value.

Rangeland: Land-use type that supplies vegetation for consumption by grazing and browsing animals. This land-use type is normally not intensively managed.

Reach: An expanse of a stream channel.

Realized Niche: Describes the part of the fundamental niche that a species actually occupies.

Recessional Moraine: Moraine that is created during a pause in the retreat of a glacier. Also called a stadial moraine.

Recharge Area: The area on the Earth's surface that receives water for storage into a particular aquifer.

Recumbent Fold: A fold in which the axial plane is almost horizontal.

Recurrence Interval: The average time period that separates natural events of a specific magnitude. For example, floods of a specific stream discharge level.

Reef: A ridge of rocks found in the tidal zone along a coastline. One common type of reef is the coral reef.

Re-Entrants: A prominent indentation in an escarpment, ridge or shoreline.

Reference Map: Map that shows natural and human-made objects from the geographical environment with an emphasis on location. Compare with thematic map.

Reflected Infrared Radiation: Form of electromagnetic radiation with a wavelength between 0.7 to 3.0 micrometers (μm).

Reflected Wave: A water wave that reflects off the shore or another obstacle and is redirected towards the sea or lake.

Reflection: Process of returning sound or light waves back to their source.

Reflection (Atmospheric): Process where insolation is redirect by 180° after striking a particle. This redirection causes 100 % loss. Most of the reflection in the Earth's atmosphere occurs in clouds because of light's interception with particles of liquid and frozen water. The reflectivity of a cloud can range from 40-90 %.

Refraction: Process where insolation is redirect to a new direction of travel after entering another medium.

Region: A term used in geography that describes an area of the Earth where some natural or human-made phenomena display similar traits.

Regional Metamorphism: Large scale metamorphic modification of existing rock through the heat and pressure of plutons created at tectonic zones of subduction.

Regolith: Loose layer of rocky material overlying bedrock.

Regosol Soil: Soil order (type) of the Canadian System of Soil Classification. This type is any young underdeveloped soil that lacks identifying soil horizons.

Relative Humidity: The ratio between the actual amount of water vapor held in the atmosphere compared to the amount required for saturation. Relative humidity is influenced by temperature and atmospheric pressure.

Relief: The range of topographic elevation within a specific area.

Remote Sensing: The gathering of information from an object or surface without direct contact.

Remote Sensor: Mechanical devices used to remotely sense an object or phenomenon.

Representative Fraction: The expression of map scale as a mathematical ratio.

Reptile: Group of terrestrial vertebrate animals that includes turtles, tortoises, snakes, lizards, crocodiles, and alligators.

Resource: Anything obtained from the environment to meet the needs of a species.

Respiration: Is the typical process where mitochondria of cells of organisms release chemical energy from sugar and other organic molecules through chemical oxidation. This process occurs in both plants and animals. In most organisms, respiration releases the energy required for all metabolic processes. This chemical reaction can be described by the following simple equation:: $C_6H_{12}O_6 + 6O_2$ >>> 6CO 2 + $6H_2O$ + released energy: Retrogressive Succession: Succession where the plant community becomes simplistic and contains fewer species and less biomass over time.

Reverse Fault: This vertical fault develops when compressional force causes the displacement of one block of rock over another.

R Horizon: Soil horizon found beneath the C horizon. Consists of consolidated rock showing little sign of weathering or pedogenesis.

Rhumb Line: A line of constant compass direction or bearing which crosses the meridians at the same angle. A part of a great circle.

Rhyolite: A fine grained extrusive igneous rock that is rich in quartz and potassium feldspar. Derived from felsic magma.

Ria Coast: An extensively carved out coast with conspicuous headlands and deep re-entrants.

Ribbon Falls: Spectacular narrow waterfalls that occur at the edge of a hanging valley.

Ribonucleic Acid (RNA): Form of nucleic acid. Ribonucleic acid is used by most organisms to read the genetic information found in DNA and to produce specific organic molecules used in the development and functioning of cells.

Richter Scale: A logarithmic measurement scale of earthquake magnitude. This scale measures the energy released by the largest seismic wave associated with the earthquake.

Riffle: Bar deposit found on the bed of streams. Associated with these deposits are pools.

Rift: Zone between two diverging tectonic plates. The mid-oceanic ridge is an area where such plate divergence is occurring.

Rift Valley: Steep sided valley found on the Earth's surface created by tectonic rifting.

Rill: A very small steep sided channel carrying water. This landscape feature is intermittent and forms for only a short period of time after a rainfall.

Rime: Deposit of ice crystals that occurs when fog or super cooled water droplets comes in contact with an object with a temperature below freezing (0° Celsius). This deposit develops outward on the windward side of the object.

Rip Current: A strong relatively narrow current of water that flows seaward against breaking waves.

Ripple: Stream bed deposit found streams. Ripples are only a few centimeters in height and spacing and are found in slow moving streams with fine textured beds.

River: A long narrow channel of water that flows as a function of gravity and elevation across the Earth's surface. Many rivers empty into lakes, seas, or oceans.

Robinson Projection: Map projection system that tries to present more accurate representations of area. Distortion is mainly manifested in terms of map direction and distance.

Roche Moutonnee: A feature of glacial erosion that resembles an asymmetrical rock mound. It is smooth and gently sloping on the side of ice advance. The lee-side of this feature is steep and jagged.

Rock: A compact and consolidated mass of mineral matter. Three types of rock are recognized: igneous, sedimentary, and metamorphic.

Rock Cycle: General model describing the geomorphic and geologic processes involved in the creation, modification and recycling of rocks.

Rockfall: Type of mass movement that involves the detachmet and movement of a small block of rock from a cliff face to its base. Normally occurs when the rock has well defined bedding planes that are exaggerated by freeze-thaw action or thermal expansion and contraction.

Salinity: Concentration of dissolved salts found in a sample of water. Measured as the total amount of dissolved salts in parts per thousand. Seawater has an average salinity of about 34 parts per thousand (ppt).

Salinization: Pedogenic process that concentrates salts at or near the soil surface because evapotranspiration greatly exceeds water inputs from precipitation.

Salt: Compounds that are produced as the result of a metal atom replacing a hydrogen atom in an acid.

Saltation: Transport of sediment initiated by moving air or water where particles move from a resting surface to the transport medium in quick continuous repeated cycles.

Salt Marsh: Coastal wetland ecosystem that is inundated for some period of time by seawater. Plants that exist in this community have special adaptation to survive in the presence of high salinities in their immediate environment. Generally, found poleward of 30° North and South latitude.

Saltwater Intrusion: The invasion of saltwater into freshwater aquifers in coastal and inland areas. This condition can be cause when groundwater, which charges the aquifer, is withdrawn faster than it is recharged by precipitation.

Sample: A sample is a subset group of data selected from a larger population group. Most samples are drawn at random to guaranty equal representation in the data.

Sand Dune: A hill or ridge of aeolian sand deposits with a minimum height of less than one meter and a maximum height of about 50 meters. Found in hot deserts and along sandy coastlines.

Sand Sea: A large region of sand and sand dunes in a desert. Common to erg deserts.

Sand Sheet: Deposit of sometimes stratified less well sorted sand that almost resemble dunes. Common in northern Europe. Believed to form when windblown materials settle on areas of patchy snow.

Sandstone: A type of sedimentary rock that contains a large quantity of weathered quartz grains.

Sand Ripples: Another term used for wind ripples.

Sand Wedge: A form of ice wedge that contains accumulations of wind blown sand in long vertical layers. A form of periglacial ground ice.

Santa Ana Wind: A warm, dry chinook like wind that occurs in southern California. Originates from the east off an elevated desert plateau.

Saturated Adiabatic Lapse Rate (SALR): The rate of decline in the temperature of a rising parcel of air after it has reached saturation. This rate is less than the dry adiabatic lapse rate (9.8° Celsius per 1000 meters) because of the heat energy added to the ascending air parcel from condensation and deposition processes.

Saturation: Atmospheric condition where water is changing its phase to liquid or solid. At saturation, relative humidity is 100 % unless there is a shortage of deposition nuclei or condensation nuclei. Generally, this process is caused by the cooling of the atmosphere.

Saturation Mixing Ratio: Mass of water vapor that a kilogram of dry air can hold at saturation. Measured in grams.

Savanna: A tropical or sub-tropical plant community

characterized by trees and shrubs scattered among a cover of grasses, herbs and forbs. The climate of a savanna is tropical with a dry season occurring in the low sun period of the year.

Scale: A specific relative or proportional size or extent of a phenomena as measured through space and/or time.

Scarification: Extensive movements of soil, sediment, and rock material caused by humans.

Scattering (Atmospheric): Is an atmospheric process where small particles and gas molecules diffuse part of the incoming solar radiation in random directions without any alteration to the wavelength of the electromagnetic energy. Scattering does, however, reduce the amount of incoming radiation reaching the Earth's surface. A significant proportion of scattered shortwave solar radiation is redirected back to space. The amount of scattering that takes place is dependent on two factors: wavelength of the incoming radiation and the size of the scattering particle or gas molecule. In the Earth's atmosphere, the presence of a large number of particles with a size of about 0.5 μm results in shorter wavelengths being preferentially scattered. This factor also causes our sky to look blue because this color corresponds to those wavelengths that are best diffused.

Schist: A medium to coarse grained metamorphic rock with well developed bedding planes derived from the foliated recrystrallization of platy like minerals like mica.

Science: Science is a way of acquiring knowledge. To do science, one must follow a specific universal methodology. The central theme in this methodology is the testing of hypotheses and the ability to make predictions. The overall goal of science is to better understand nature and our Universe.

Scientific Method: The approach science uses to gain knowledge. This method tries to be unbias and neutral. Involves inductive and deductive reasoning, hypothesis testing and falsification, and predictive model testing.

Sclerophyllous Vegetation: Term used to describe drought resistant vegetation common in Mediterranean climates. Some common adaptations present in this type of vegetation include: deep roots, reduced leaf area exposed to the atmosphere, and waxy thick leaves with closing stomata which resist water loss.

Scour: Abrasive effects of rocks and sediments incorporated in the ice base of a glacier.

Scree: An accumulation of weathered rock fragments at the base of a steep rock slope or cliff.

Sea: A body of saline water found on the Earth's continental surface. A portion of a ocean that is in close proximity to a continent.

Sea Arch: A coastal landform composed of rock that resembles an arch. These landforms are created when waves erode through a thin headland from both sides.

Sea Breeze: Local thermal circulation pattern found at the interface between land and water. In this circulation system, surface winds blow from water to land during the daytime.

Sea-Floor Spreading: The process of oceanic crust creation and sea-floor movement that occurs at the mid-oceanic ridge.

Sea-Level: The average surface elevation of the world's oceans.

Sea-Level Pressure: Average atmospheric pressure at sea-level. This value is 1013.2 millibars.

Seamount: A volcanic mountain found on an ocean basin that has an origin not related to a mid-oceanic ridge or a tectonic subduction zone.

Seasons: Time periods generally based on the changes in the intensity and duration of sunlight as received in the middle and high latitudes. Four seasons are normally recognized: Spring; Summer; Fall; and Winter. The astronomical definition is more precise and suggests the following time periods for the four seasons: Spring - March 22 to June 21; Summer - June 22 to September 22; Fall - September 23 to December 22; and Winter - December 23 to March 21.

Sea Stack: A steep pillar of rock located in the ocean a short distance from the coastline. These landforms are created when waves erode through a thin headland from both sides.

Seaward: Positioned or located away from land but towards an ocean or sea.

Seawater: The mixture of water and various dissolved salts found in the world's oceans and seas.

Secondary Consumer: Organisms that occupy the third trophic level in the grazing food chain. These organisms are carnivores.Also known as a primary carnivore.

Secondary Substance: Organic chemical produced by a plant that has no direct function in its metabolism. Many of these chemicals are toxic and are believed to be created by the plant to reduce herbivore damage or the negative effects of competition by other plants.

Secondary Succession: Succession on a previously vegetated surface. The soil or substrate on this location contains an active seed bank.

Second-Growth Forest: Stand of forest that is the result of secondary succession.

Second Law of Thermodynamics: This law states that heat can never pass spontaneously from a colder to a hotter body. As a result of this fact, natural processes that involve energy transfer must have one direction, and all natural processes are irreversible. This law also predicts that the entropy of an isolated system always increases with time.

Seismic Wave: Successive wave-type displacement of rock usually caused by an earthquake.

Seismograph: Instrument that measures the energy contained in seismic waves from an earthquake or other type of ground displacement.

Seismology: A branch of science focused on the study of earthquakes and seismic activity.

Self-Regulation: The ability of some systems to maintain a steady state equilibrium through positive and negative feedbacks.

Semi-Diurnal Tide: Tides that have two high and two low waters per tidal period.

Sensible Heat: Heat that can be measured by a thermometer and thus sensed by humans.

Sensible Heat Flux: Process where excess heat energy is transferred into the atmosphere. The process first involves the movement of heat energy from the Earth's surface to the atmosphere by conduction and convection. The heat energy then can move horizontally advection (atmospheric circulation).

Sexual Reproduction: Any process of reproduction that does involve the fusion of gametes.

Shale: Fine grained sedimentary rock composed of lithified clay particles.

Shear Stress: Stress caused by forces operating parallel to each other but in opposite directions.

Shear Wave: A seismic wave that creates wave-like motion perpendicular to the direction of seismic energy propagation. Also called S-wave.

Shore: The land area bordering a relatively large water body like a lake or ocean.

Short Wave: A small wave in the polar jet stream and the westerlies that extends from the middle to the upper troposphere. Often associated with the formation of a mid-latitud cyclone at the ground surface. Contrasts with long waves.

Shortwave Radiation: Electromagnetic radiation with a wavelength between 0.1 and 0.7 micrometers (μm). Commonly used to describe the radiation emitted from the sun.

Shrub: A woody plant species that is smaller than a tree. Shrubs usually do not have a trunk.

Sial Layer: The part of the crust that forms the continents and is composed of relatively light, granitic rocks.

Siberian High: High pressure system that develops in winter over northern central Asia.

Sidereal Day: Time it takes to complete one Earth rotation relative to the position of a fixed star. This measurement takes 23 hours, 56 minutes, and 4.09 seconds. Compare with mean solar day.

Silicate: Group of minerals that have crystal structures based on a silica tetrahedron (SiO_4).

Silicate Magma: Magma that is felsic in composition.

Silcretes: Sedimentary rock created by the chemical precipitation of silica.

Siltstone: Fine grained sedimentary rock composed of lithified silt particles.

Silurian: Geologic period that occurred roughly 408 to 438 million years ago. During this period, the first plant and insect species appear.

Sima Layer: The part of the crust that forms the ocean basins and lower layers in the crust and is composed of relatively heavy, basaltic rocks.

Sinkhole: A pit like hole in found in areas of karst. These features are caused by the weathering of limestone or dolomite by subsurface drainage. Also called a sink or *doline*.

Sinusoidal Equal-Area Projection: Map projection that represents areas in their true form on a two-dimensional map. Distances are only correct along parallels and central meridian. Shapes become more distorted away from the central meridian and close to the poles.

Slate: A fine grained metamorphic rock with well developed bedding planes derived from the sligt recrystrallization of shale.

Sling Psychrometer: Psychrometer that uses a rotating handle and a whirling motion to ventilate its wet-bulb thermometer.

Slip-Face: The lee side of a dune where material accumulates and slides or rolls downslope.

Slope Aspect: Main compass direction (North, North East, East, South East, South, South West, West, and North West) that a slope faces.

Slope Failure: The downslope movement of soil and sediment by processes of mass movement.

Small Circle: A circle on the globe's surface that does not bisect the center of the Earth. Parallels of latitude are examples of small circles.

Snout: Front end of a glacier. Also called the terminus.

Snow: A type of solid precipitation that forms in clouds with an air temperature below freezing. Snow forms when water vapor deposits directly as a solid on a deposition nuclei. Snowflakes begin their life as very tiny crystals developing on a six-sided hexagonal deposition nuclei. The developing snowflak, then grows fastest at the six points of the nuclei as these surfaces are more exposed to atmosphere's water vapor. Snowfall is most common with the frontal lifting associated with mid-latitude cyclones during fall, winter, and spring months when air temperatures are below freezing.

Snowfield: An area of permanent snow accumulation. Usually at high altitudes or latitudes.

Snow Line: Altitudinal or latitudinal limit separating zones where snow does not melt during the summer season from areas in which it does. Similar to the concept of firm limit except that it is not limited to glaciers.

Snow Melt: Conversion of snow into runoff and groundwater flow with the onset of warmer temperatures.

Snow Pellets: A form of precipitation also known as graupel. Snow pellets are white, spherical bits of ice with a maximum diameter of 5 millimeters. Snow pellets develop when supercooled droplets freeze on snowflakes. Snow pellets often fall for a brief time period when precipitation transforms from ice pellets to snow. Snow pellets can be easily distinguished from packed snowflakes as they tend to bounce when they strike the ground. Packed snowflakes are not dense enough to cause them to bounce.

Soil: Layer of unconsolidated material found at the Earth's surface that has been influenced by the soil forming factors: climate, relief, parent material, time, and organisms. Soil normally consists of weathered mineral particles, dead and living organic matter, air space, and the soil solution.

Soil Colloids: Very small organic and inorganic particles found in a soil. Inorganic colloids are often clay particles. Soil colloids carry a negatve electrical charge and are the primary sites for cation exchange. Soil colloids hold large quantities of elements and compounds which are used by plants for nutrition.

Soil Creep: Slow mass movement of soil downslope. Occurs where the stresses on the slope material are too small to create a rapid failure.

Soil Erosion: Transport of soil mineral particles and organic matter by wind, flowing water, or both. Human activities that disturb the soil surface or remove vegetation can enhance this natural process.

Soil Fertility: The ability of a soil to provide nutrients for plant growth.

Soil-Heat Flux: The rate of flow of heat energy into, from, or through the soil.

Soil Horizon: Layer within a soil profile that differs physically, biologically or chemically from layers above and/or below it.

Soil Moisture Recharge: The process of water filling the pore space found in a soil (storage).

Soil Permeability: The rate at which water and air move vertically through a soil.

Soil Porosity: The volume of water that can be held in a soil. Also refers to the ratio of the volume of voids to the total volume of the soil.

Soil Profile: Vertical arrangement of layers or horizons in a soil.

Soil Science: The study of soils from an interdisciplinary perspective.

Soil Solution: Aqueous liquid found within a soil. This liquid normally contains ions released from mineral particles, organic matter or plant roots and leaves.

Soil Structure: General term that describes how mineral and particles organic matter of are organized and clumped together in a soil.

Soil Taxonomy: The classification of a soil in a hierarchical system based on its various properties. Grouping goes from general to specific.

Soil Texture: The relative quantities of the different types and sizes of mineral particles in a soil.

Soil Water: The water found occupying the pore spaces between soil particles.

Solar Altitude: Height of the sun above the horizon from either True North or True South.

Solar Constant: A term used to describe the average quantity of solar insolation received by a horizontal surface at the edge of the Earth's atmosphere. This value is approximately 1370 Watts per squar meter.

Solar Day: Time required for the Earth to complete one rotation relative to the sun.

Solar Noon: Point of time during the day when the sun is aligned with True North and True South.

Solar System: The collection of celestial bodies that orbit around the sun.

Solar Wind: Mass of ionized gas emitted to space by the sun. Plays a role in the formation of auroras.

Solar Year: The time it takes the Earth to make one orbit around the sun. This is approximately 365.2422 days.

Solid: A state of matter where molecules where the mass of the substance does not have the property of flow.

Solonetzic Soil: Soil order (type) of the Canadian System of Soil Classification. This soil type is associated with grassland habitats where high levels of evapotranspiration cause the accumulation of salts at or near the soil surface. These soils are common in the dry regions of the prairies where evapotranspiration greatly exceeds precipitation input.

Solstice: Dates when the declination of the sun is at 23.5° North or South of the equator. For the Northern Hemisphere this date falls on June 21 or 22 (Summer Solstice). In the Southern Hemisphere the date is December 21 or 22 (Winter Solstice).

Solum: Part of the soil that is capable of supporting life.

Solution: Form of chemical weathering where rocks and minerals are dissolved by water. Materials entering the mixture can alter the chemical nature of the solution and can increase the strength of this weathering agent. For example, the mixing of carbon dioxide and water can form carbonic acid. The dissolving of a substance into a liquid.

Source Region: Area where air masses originate and come to possess their moisture and temperature characteristics.

Southern Oscillation: Reversal of atmospheric circulation in tropical Pacific Ocean that triggers the development of an El Nino.

South Magnetic Pole: Location in the Southern Hemisphere where the lines of force from Earth's magnetic field are vrtical. This point on the Earth gradual changes its position with time.

South Pole: Surface location defined by the intersection of the polar axis with Earth's surface in the Southern Hemisphere. This location has a latitude of 90° South.

Spatial Analysis: The examination of the spatial pattern of natural and human-made phenomena using numerical analysis and statistics.

Spatial Isolation: Reproductive isolation of two or more populations of a species by distance or physical barriers. Over long periods of time geographic isolation leads to speciation through divergent evolution because of environmental heterogeneity. Also called geographic isolation.

Spatial Tradition: Academic tradition in modern Geography that investigates geographic phenomena from a strictly spatial perspective.

Specialist Species: Species that have a relatively narrow ecological niche. These species may be able to live in only one type of habitat, tolerate only specific environmental conditions, or use only a few types of food.

Speciation: The process by which new species originate through mutations, natural selection, and evolution.

Taku: Name for a katabatic type of cold wind that occurs in Alaska.

Talik: An unfrozen section of ground found above, below, or within a layer of discontinuous permafrost. These layers can also be found beneath water bodies in a layer of continuous permafrost. A number of different types of talik have been distinguished: closed talik, open talik, and through talik.

Talus: An accumulation of angular rock debris from rockfalls.

Talus Slope: A slope that is composed of talus.

Tarn: A small mountain lake that occurs inside a cirque basin.

Taxon: A classification category for a group of organisms.

Taxonomic Classification: Classification of organisms based on structural and physiological connections between other species.

Tectonic Plate: An extensive layer of lithosphere that moves as a discrete unit on the surface of the Earth's asthenosphere.

Temperate Deciduous Forest: Forested biome found in the mid-latitudes and dominated by deciduous vegetation.

Temperate Glacier: Glacier in which the ice found below 10 to 20 meters from its surface is at the pressure melting point. One of the three types of glaciers: cold glacier; temperate glacier; and subpolar glacier.

Temperate Rain Forest: An ecosystem that is dominated by large and very tall evergreen trees. This biome occurs along the Pacific Northwest coast of North America where annual precipitation is high and temperatures are mild.

Temperature: Temperature is defined as the measure of the average speed of atoms and molecules. The higher the temperature the faster they move.

Temperature Inversion: Situation where a layer of warmer air exists above the Earth's surface in a normal atmosphere where air temperature decreases with altitude. In the warmer layer of air, temperature increases with altitude.

Tephra: Fragmented rock material ejected by a volcanic explosion. Also called *pyroclastic material*.

Terminal Fall Velocity: Velocity at which a particle being transported by wind or water falls out of the moving medium. This velocity is dependent on the size of the particle.

Terminal Moraine: Moraine that marks the maximum advance of a glacier.

Terminal Velocity: Maximum speed that can be achieve by a body falling through a fluid like water or air.

Terrace: An elevated surface above the existing level of a floodplain or shore that is created by stream or ocean wave erosion.

Tertiary: Geologic period that occurred roughly 1.6 to 65 million years ago. During this period, mammals become a dominant species on the planet.

Tertiary Consumer: Organisms that occupy the fourth trophic level in the grazing food chain. These organisms are carnivores. Also known as a secondary carnivore.

Tetrahedron: Silicon atom joined by four oxygen atoms (SiO_4). The atomic properties of this molecule cause it to develop a unique three dimensional crystal lattice that is pyramid shaped.

Thematic Map: Map that displays the geographical distribution of one phenomenon or the spatial associations that occur between a few phenomena. Compare with reference map.

Thematic Mapper: Remote sensing device found on Landsat satellites that scans images in seven spectral bands from visible to thermal infrared.

Thermal Circulation: Atmospheric circulation caused by the heating and coling of air.

Thermal Equator: Continuous area on the globe that has the highest surface temperatures because of the presence of the Intertropical Convergence Zone.

Thermal High: Area of low pressure in the atmosphere caused by surface temperatures.

Thermal Infrared Radiation: Form of electromagnetic

radiation with a wavelength between 3 to 14 micrometers (μm).

Thermal Low: Area of high pressure in the atmosphere caused by surface temperatures.

Thermal Metamorphism: Is the metamorphic alteration of rock because of intense heat released from processes related to plate tectonics.

Thermocline: Boundary in a body of water where the greatest vertical change in temperature occurs. This boundary is usually the transition zone between the layer of warm water near the surface that is mixed and the cold deep water layer.

Thermodynamic Equilibrium: This type of equilibrium describes a condition in a system where the distribution of mass and energy moves towards maximum entropy.

Thermodynamic Laws: Laws that describe the physical processes, relationships, and phenomena associated with heat.

Thermokarst: Landscape dominated by depressions, pits, and caves that is created by the thawing of ground ice in high latitude locations. Resembles karst landscape but is not created by chemical weathering.

Thermometer: Device used to measure temperature.

Thermosphere: Atmospheric layer above the mesosphere (above 80 kilometers) characterized by air temperatures rising rapidly with height. The thermosphere is the hottest layer in the atmosphere. In the thermosphere, gamma, X-ray, and specific wavelengths of ultraviolet radiation are absorbed by certain gases in the atmosphere. The absorbed radiation is then converted into heat energy. Temperatures in this layer can get as high as 1300-1800° Celsius.

Third Law of Thermodynamics: This law states if all the *thermal motion* of molecules (kinetic energy) could be removed, a state called absolute zero would result and all energy would be randomly distributed.

Threatened Species: Species that is still plentiful in its natural range but is likely to become endangered because of declining population numbers.

Threshold: The level of magnitude of a system process at which sudden or rapid change occurs.

Throughflow: The roughly horizontal flow of water through soil or regolith.

Through Talik: Is a form of localized unfrozen ground (talik) in an area of permafrost. It is open to the ground surface and to an area of unfrozen ground beneath it. Permafrost encases it along the sides.

Thrust Fault: A geologic fault where the hanging wall is forced over the foot wall.

Thunder: Sound created when lightning causes the rapid expansion of atmospheric gases along its strike path.

Thunderstorm: A storm several kilometers in diameter created by the rapid lifting of moist warm air which creates a cumulonimbus cloud. Thunderstorms can have the following severe weather associated with them: strong winds; hail; lightning; tornadoes; thunder; and heavy rain.

Tidal Current: Regional scale ocean current that is created the tidal rise and fall of the ocean surface.

Tidal Period: Time it takes for one tidal cycle.

Tidal Zone: Area along the coastline that is influence by the rise and fall of tides.

Tide: Cyclical rise and fall of the surface of the oceans. Caused by the gravitational attraction of the sun and moon on the Earth.

Till: Heterogeneous sediment deposited directly by a glacier. The particles within this deposit have not been size sorted by the action of wind or water.

Till Plain: Extensive flat plain of till that forms when a sheet of ice becomes detached from the main body of the glacier and melts in place depositing the sediments it carried.

Time: Measurable period in which cause and effect occurs and systems function.

TIROS (Television and Infrared Observation Satellite):

Series of meteorological satellites launched by the United States starting in 1960. The main purpose behind these satellites was to use a variety of remote sensing devices for weather forecasting. TIROS program was very successful, providing the first accurate weather forecasts based on data gathered from space. TIROS began continuous monitoring of the Earth's weather in 1962.

Tissue: A group of similar cells that are organized into a structure with a specific purpose.

Tolerance Model of Succession: This model of succession suggests that the change in plant species dominance over time is caused by competition for resources. Later species are able to tolerate lower resource levels due to competition and can grow to maturity in the presence of early species, eventually out competing them.

Tolerance Range: Limits of tolerance a species has to an abiotic factor or condition in the environment.

Tombolo: A coastal feature that forms when a belt sand and/ or gravel is deposited between an island and the mainland. This feature is above sea-level for most of the time.

Topographic Map: Map that displays topography through the use of elevation contour lines. Base elevation on these maps is usually sea-level.

Topographic Profile: A two-dimensional diagram that describes the landscape in vertical cross-section.

Topography: The relief exhibited by a surface.

Topset Bed: Horizontal deltaic deposit composed of coarse alluvial sediment. Represents current or past surface of the delta.

Tornado: A vortex of rapidly moving air associated with some severe thunderstorms. Winds within the tornado funnel may exceed 500 kilometers per hour.

Tornado Alley: Region in North America which receives a extraordinary high number of tornadoes. This region stretches from central Texas to Illinois and Indiana.

Tornado Warning: A warning issued to the public that a

tornado has been observed by an individual in a specified region. This warning can also be issued if meteorological information indicates a high probability that a tornado will develop in a specified region.

Tornado Watch: A forecast issued to the public that a tornado may occur in a specified region.

Total Column Ozone: A measurement of ozone concentration in the atmosphere.

Traction: Erosional movement of particles by rolling, sliding and shuffling along the eroded surface. Occurs in all erosional mediums (air, water, and ice).

Trade Winds: Surface winds that generally dominate air flow in the tropics. These winds blow from about 30° North and South latitude (subtropical high pressure zone) to the equator (intertropical convergence zone). Trade winds in the Northern Hemisphere have northeast to southwest direction and are referred to as the *Northeast Trades*. Southern Hemisphere trade winds have southeast to northwest direction but are called the *Southeast Trades*.

Transform Fault: Massive strike-slip fault continental in size. Examples of such faults occur along tectonic plate boundaries and at the mid-oceanic ridge.

ltisols: Soil order (type) of the United States Natural Resources Conservation Service Soil Classification System. Tropical or subtropical soil that is in an advanced state of development. These soils have pronounced eluviation, clay accumulation in a subsurface layer, and are often poor in base cations.

Ultramafic: Rock that is rich in magnesium and iron content.

Ultraviolet Radiation: Electromagnetic radiation with a wavelength between 0.1 and 0.4 micrometers (μm).

Unconfined Aquifer: Aquifer that is not restricted by impervious layers of rock.

Unconfined Groundwater: Groundwater that is not restricted by impervious layers of rock.

Unconformity: A break in the sequence of sedimentary strata. Often the unconformity surface is the result of erosion.

Undercut Bank: Steep bank found on the inside of stream meanders. Formed by the erosion that occurs when a stream channel moves horizontally.

Uniformitarianism: Is a theory that rejects the idea that catastrophic forces were responsible for the current conditions on the Earth. The theory suggested instead, that continuing uniformity of existing processes were responsible for the present and past conditions of this planet.

United States Natural Resources Conservation Service Soil Classification System: A hierarchical system that is used in United States to classify soils. This system has six levels: *order*, *suborder*, *great group*, *subgroup*, *family*, and *series*. At the *order* level, 12 types of soils are recognized: gelisols, oxisols, aridsols, mollisols, alfisols, ultisols, spodsols, entisols, inceptisols, vertisols, histosols, and andisols.

Universal Time (UT): The mean solar time of the meridian at the Prime Meridian. Universal Time replaced the time standard known as Greenwich Mean Time in 1928. Universal Time is commonly used to denote solar time.

Universal Transverse Mercator (UTM) Grid System: Rectangular coordinate system used to find location of points on the Earth's surface. Based on the Universal Transverse Mercator projection system.

Universe: All of the observable phenomena in the *celestial cosmos*.

Unloading: The releasing of downward pressure on rocks because of removal of overlying material by erosion. Unloading can cause the development of horizontal bedding in once solid rock.

Unstable Atmosphere: Condition in the atmosphere where isolated air parcels have a tendency to rise. The parcels of air tend to be warmer than the air that surrounds them.

Unstable Equilibrium: In an unstable equilibrium the system returns to a new equilibrium after distrbance.

Updraft: Upward movement of air.

Upper Air Westerlies: Consistent winds that exist in the upper troposphere that flow east to west from about 20° of latitude to the poles.

Upper Mantle: Layer of the Earth's interior extending from the base of the crust to 670 kilometers below the surface. Part of the Earth's mantle layer. The upper mantle is composed of peridotite, an ultramafic magma primarily made up of the minerals olivine and pyroxene. The top layer of the upper mantle, 100-350 km below surface, is called the asthenosphere.

Upslope Fog: Fog produced by air flowing over topographic barriers. As the air is forced to rise, it is cooled by adiabatic expansion. Upslope fog is most common on the windward slopes of hills or mountains.

Upwelling: The movement of nutrient-rich deep seawater to the ocean's surface.

Valley: A linear depression in the landscape that slopes down to a stream, lake or the ocean. Formed by water and/or ice erosion.

Valley Breeze: Local thermal circulation pattern found in areas of topographic relief. In this circulation system, surface winds blow from the valley bottom to areas of higher elevation during the daytime.

Valley Fog: Fog formed by the movement of cooler, more dense air from higher elevations to the warm valley bottom.

Valley Train: A linear accumulation of glaciofluvial outwash sediments found in a once glaciated valley.

Valley Wall: The side slope of a stream or glacial valley.

Vapor Pressure: Pressure exerted by water vapor molecules in a given quantity of atmosphere.

Variance: A statistical measure of the dispersion of observation values in a data set. The variance of a sample is the sum of the square of each value in the data set subtracted from the mean divided by one less than the total number of observations in the data set.

Varve: A thin yearly deposit of sediment found on the bottom of a lake. Within each yearly varve, there are variations in the color and the texture of the material deposited. The thickness of the varve and its associated layers can be used to reconstruct past environmental conditions influencing the lake.

Vascular Plant: Plant that has vascular tissues to transport water, nutrients, and other metabolic products.

Velocity: The speed of movement of an object in one direction.

Ventifact: A loose piece of rock that has been polished smooth by wind transported particles. Common in arid environments.

Venturi: An increase in the velocity of a fluid or gas due to the constriction of flow.

Vernal Equinox: One of the two periods when the declination of the sun is at the equator. The vernal equinox occurs on March 20 or 21.

Vertical Aerial Photograph: Photograph taken from a overhead or near overhead angle from a platform in the atmosphere.

Warm Desert: Desert found in the subtropics or interiors of continents at the middle latitudes where precipitation is low and surface air temperatures are high.

Warm Front: A transition zone in the atmosphere where an advancing warm air mass displaces a cold air mass.

Water Consumption: The complete removal of water from some type of source, like groundwater, for some use by humans. This water is not returned to the source. Compare with water withdrawal.

Waterfall: A location in the long profile of a stream where water flows vertically. A nickpoint. Verical drop in elevation that causes a stream's dischange to flow vertically.

Watershed: Catchment area of a drainage basin.

Waterspout: A vortex of rapidly moving air over water that is associated with some thunderstorms.

Water Table: Top surface of groundwater.

Water Withdrawal: The removal of water from some type of source, like groundwater, for some use by humans. The water is subsequently returned some period of time later after its is used. The quality of the returned water may not be the same as when it was originally removed. Compare with water consumption.

Watt: A metric unit of measurement of the intensity of radiation in Watts over a square meter surface (W/m^2 or $W\ m^{-2}$).

Wave: A moving swell or ridge on the surface of a solid or liquid or within the medium of a gas. Electromagnetic radiation also travels in waves.

Wave Crest: The curved tops or ridges of an oscillating wave.

Wave-Cut Notch: A rock recess at the foot of a sea cliff where the energy of water waves is concentrated.

Wave-Cut Platform: A flat or slightly sloping bedrock surface that forms in the tidal zone. Caused by wave erosion.

Wave Height: Vertical distance between a wave's trough and crest.

Wavelength: Distance beween two successive wave crests or troughs.

Wave Period: The time elapsed for a wave to travel the distance of one wavelength.

Wave Refraction: The re-orientation of a wave so that it approaches a shoreline at a more perpendicular angle. This process is caused by the differential reduction of water depth as a linear wave approaches a curved shoreline. A reduction in water depth causes a wave to slow down causing the waves approaching a nonlinear shoreline to curve with the shore's shape.

Wave Trough: Area in between wave crests.

Weather: The state of the atmosphere at a specific time and place.

Weathering: Physical, chemical or biological breakdown of rocks and minerals into smaller sized particles.

Weathering Landform: Is a landform created by the physical or chemical decomposition of rock through weathering. Weathering produces landforms where rocks and sediments are decomposed and disintegrated. This includes landforms with some of the following geomorphic features: karst, patterned ground, and soil profiles.

Weather Map: Map that displays the condition of the physical state of the atmosphere and its circulation at a specific time over a region of the Earth.

Westerlies: Dominant winds of the mid-latitudes. These winds move from the subtropical highs to the subpolar lows from west to east.

Wet-Bulb Depression: The value calculated by subtracting a wet-bulb thermometer reading from a dry-bulb thermometer reading. Used to determine the air's relative humidity or dew point from a psychrometric table.

Wet-Bulb Thermometer: Thermometer on a psychrometer that has a moisten wick on its reservoir bulb. When ventilated this thermometer records a temperature that is modified by the cooling effects of evaporation. This measurement and the temperature reading from a dry-bulb thermometer are then used to determine the air's relative humidity or dew point from a psychrometric table.

Wet Deposition: The transport of gases and minute liquid and solid particles from the atmosphere to the ground surface with the aid of precipitation or fog. Compare with dry deposition.

Wetland: Natural land-use type that is covered by salt water or fresh water for some time period. This land type can be identified by the presence of particular plant species or characteristic conditions.

Wetting and Drying: Physical weathering process where rocks are mechanically disintegrated by the accumulation of successive layers of water molecules in between the mineral grains of a rock. Sometimes called slaking.

Wien's Law: This radiation law suggests that the wavelength of maximum emission of any body is inversely proportional to its absolue temperature. The following equation mathematically describes this law:: lmax = C/T: where lmax is the body's maximum emitted wavelength of radiation in micrometers (μm),: C is a constant equal to 0.2897,: and T is the temperature of the body in Kelvins.

Wilting Point: The point at which the rate of water leaving a plant's leaves is greater than the water uptake by the roots. At this point the plant will fail to recover its turgidity.

Wind: Air moving horizontally and/or vertically.

Windward: Upwind side or side directly influenced to the direction that the wind blows from. Opposite of leeward.

Winter: Season between fall and spring. Astronomically it is the period from the winter solstice to the vernal equinox in the Northern Hemisphere.

Winter Solstice: Date when the declination of the sun is at 23.5° South of the equator. This date is usually December 21 or 22.

X-axis: Horizontal axis on a graph.

Xerophyte: Plant that have adaptations to survive prolonged periods of soil drought.

X-Ray Radiation: Form of electromagnetic radiation with a wavelength between 0.03 to 30 nanometers.

Xylem: Conducting tissue in vascular plants through which water and mineral nutrients are transported.

Yardang: Rock that has developed a streamline form because of wind erosion. The long axis of these features is aligned with the dominant wind direction.

Yazoo Tributary: Small tributary channel that is prevented from joining the main stream channel by the presence of levees. Yazoo tributaries tend to flow on the floodplain parallel to the main stream channel.

Younger-Dryas: A cold period during the generally mild Holocene epoch that occurred from about 10,000 - 8,500 BC. Scientists speculate that this cooling may have been caused by the release of fresh water trapped behind ice on North America into the North Atlantic Ocean.

Zonal: Movement of wind or ocean waters in a direction that is roughly parallel to the lines of latitude.

Zone of Ablation: Area of a glacier where losses of ice from melting, evaporation, and sublimation exceed additions of snow annually.

Zone of Accumulation: Area of a glacier where additions of snow exceed losses of ice from melting, evaporation, and sublimation.

Zone of Aeration: Horizontal zone that extends from the top of the water table to the ground surface. Soil and rock pore spaces in this zone may and may not have water.

Zone of Saturation: Groundwater zone within the Earth's bedrock where all available pores spaces are filled by water. Found beneath the water table.

Bibliography

Allaby, M. 1998. Oxford Dictionary of Ecology. Oxford University Press, New York, NY.

Arthur Lovejoy, *The Great Chain of Being* (Harvard University Press, 1936).

Athanasius Kircher, *Mundus Subterraneus* (Amsterdam, 1665). See Marjorie Nicolson, *Mountain Gloom and Mountain Glory* (New York: Cornell University Press, 1959), pp. 168-73.

Attrill, M.J. and S.D. Rundle. 2002. Ecotone or ecocline: ecological boundaries in estuaries. Estuarine, Coastal, and Shelf Science 55: 929-936.

B. Moss, *Ecology of Fresh Waters* (Blackwell, 1998).

Bacon, 'De Dignitate et Augmentis Scientarum' in *Works,* ed. J. Spedding, R. Ellis. D. Heath (London: Longmans, 1870). vol. 4, p. 296.

Berdoulay, V. (1981), The contextual approach, in D. R. Stoddart (ed) Geography, ideology and social concern. Black well, Oxford.

Bernhard Edmaier, *Earthsong*. A collection of breathtaking arial photographs. Phaidon, 2004.

Boyce, S.G. 1995. Landscape Forestry. John Wiley and Sons, Inc., New York, NY.

Bradley, Raymond S. (1985.) *Quaternary paleoclimatology : methods of paleoclimatic reconstruction* (Boston: Allen & Unwin).

Brookfield H.C. (1962) Local study and comparative method. *Annals of the Association of American Geographers*, 52, 242-54.

Brookfield. H.C. (1968) The money that grows on trees, *Australian Geographical Studies,* 6, 97-119.

Bunge, W. (1971) *Fitzgerald: geography of a revolution.* Schenkman, Combridge, Mass.

Buol, Stanley W., F,D. Hole and R.W. McCracken. 1997. Soil Genesis and Classification, 4th ed. Iowa State Univ. Press.

Burgess, K. (1984) *In the Field: an introduction to field research.* Allen &. Unwin, London.

Carolyn Merchant, *The Death of Nature* (London: Wildwood, 1982), 'Organic Society and Utopia', pp. 69-99.

Clarke. C., Ley, D. and Peach C. (eds.) (1984), *Geography and Ethnic Pluralism.* Allen and Unwin, London.

Companion to Feminist Geography (Blackwell Companions to Geography), ed. by Joni Seager, Lise Nelson, Blackwell Publishers 2004.

Crowley, Thomas J., and North, Gerald R. (1996) *Paleoclimatology* (Oxford, UK: Clarendon Press) Series : Oxford monographs on geology and geophysics no. 18.

Dangerfield, J.M., A.J. Pik, D.Britton, A. Holmes, M. Gillings, I. Oliver, D. Briscoe, and A.J. Beattie. 2003. Patterns of invertebrate biodiversity across a natural edge. Austral Ecology 28: 227-236.

Edwin Arthur Burtt, *The Metaphysical Foundations of Modern Physical Science (London:* Kegan Paul, 1925); ch. 3: 'Galileo', pp. 61-95.

Eyles, J., 1986, The geography of everyday life, in D. Gregory & R. Walford (eds) New horizons in human geography, Macmillan, London.

Forman, R.T.T. 1995. Land Mosaics: The Ecology of Landscapes and Regions. Cambridge University Press, Cambridge, UK.

Forman, R.T.T. and M. Godron. 1986. Landscape Ecology. John Wiley and Sons, Inc., New York, NY, USA.

G.E. Hutchinson, *A Treatise on Limnology*, 3 vols. (1957-1975) - classic but dated.

Gerald A. Cole, *Textbook of Limnology*, 4th ed. (Waveland Press, 1994)

Gould, J. and Kolb, W.L., (Eds.), (1964), *A Dictionary of the Social Sciences* (Paris).

Gould, Stephen Jay. (c1989) *Wonderful life, the story of the Burgess Shale* (New York: W.W. Norton).

H.B.N. Hynes, *The Ecology of Running Waters* (1970).

Haggett, P. (1965), *Locational Analysis in Human Geography,* (London).

Haslinga, N.W. (1983), Between German and French geography, Tijdschrift Voor Economicshe en Sociale Geography 74, 317-34.

Hole Francis D., J.B. Campbell. 1985. Soil landscape analysis. Totowa Rowman & Allanheld, p. 214.

Imbrie, John. (1986) *Ice ages : solving the mystery* (Cambridge, Massachusetts: Harvard University Press, 1986, c1979).

Jackson, P. S. Smith (1934), Exploring social geography. Alien and Unwind, London.

Jacob Kalff, *Limnology* (Prentice Hall, 2001).

John Barrell, *The Dark Side of the Landscape: The Rural Poor in English Painting 1730-1840* (Cambridge University Press, 1980).

Johnston, R.J., and P. Caval (1984), Geography since the second world war. Croom Helm, London.

Kimble, G.H.T. (1951): The Inadequacy of the Regional Concept, in Essays in Geography, edts. L.D.Stamp and S.W. Wooldridge, London, pp. 1951-174.

Ley, D. (1920) Liberal ideology and the post-industrial city, *Annals of the Association of American Geographers*. 70, 238-58.

Ley, D. (1977) Social geography and the taken for granted world, *Transactions of the Institute of British Geographers,* NS 2, 498-512.

Linda Mcdowell. *Gender, Identity and Place: understanding feminist geographies*, Cambridge : Polity Press, 1999--introduction

Loan, L. Van dread A. Piersma (1982), The image of man: paradigmatic cornerstone in human geography, Annals, Associations of American Geographers 72, 411-26.

Loffler, J. and O.-D. Finch. 2005. Spatio-temporal gradients between high mountain ecosystems of central Norway. Arctic, Antarctic, and Alpine Research 37: 499-513.

Lucile Brockway, *Science and Colonial Expansion: The Role of the British Royal Botanic Gardens* (London: Academic Press, 1979), pp. 141-65.

Lyon, J. and C. L. Sagers, C.L. 1998. Structure of herbaceous plant assemblages in a forested riparian landscape. Plant Ecology 138: 1-16.

M.H. Abrams. *The Mirror and the Lamp* (Oxford University Press, 1953). pp. 298-335.

M.J. Selby, *Earth's Changing Surface*, Oxford University Press, 1985

MacArthur, Robert H. and Wilson, Edward O. *The Theory of Island Biogeography* Princeton University Press. 2001.

MacLeod, Gordon and Jones, Martin (2001): Renewing The Geography if Regions. Environment and Planning D 16(9), pp. 669-695.

Magnuson, J.J. 1991. Fish and fisheries ecology. Ecological Applications 1: 13-26.

Malczewski, J. 1999. GIS and Multicriteria Decision Analysis. John Wiley and Sons, Inc., New York, NY, USA.

Margulis, Lynn, and Dorion Sagan. (c1986) *Origins of sex: three billion years of genetic recombination* (New Haven : Yale University Press) Series : The Bio-origins series.

Marjorie Nicolson, *Newton Demands the Muse* (Princeton University Press, 1946).

Merchant, *The Death of Nature;* Susan Griffin, *Woman and Nature* (New York: Harper and Row, 1978).

Miriam Rothschild, *Dear Lord Rothschild* (London: Hutchinson, 1983).

Naveh, Z. and A. Lieberman. 1984. Landscape ecology: theory and application. Springer-Verlag, New York, NY, USA.

Niavis, Judicum Jovis (Leipzig, n.d.). For a fuller account of this allegory, see Frank Dawson Adams, *The Birth and Development of the Geological Sciences* (New York: Dover, 1938), pp. 171-5.

Pliny, *Natural History,* trans. J. Bostock and H. T. Riley (London: Bohn, 1858), vol. 6, Bk 33, ch. 1.

R.M. Young, 'Malthus and the Evolutionists: The Common Context of Biological and Social Theory', *Past and Present,* 43 (1969), pp. 109-45.

R.M. Young, 'The Historiographic and Ideological Contexts of the Nineteenth Century Debates on Man's Place in Nature', in M. Teich and R. Young (eds.). *Changing Perspectives in the History of Science* (London: Heinemann, 1973), pp. 344-438.

R.M. Young, 'The Human Limits of Nature', in J. Benthall (ed.). *The Limits of Human Nature (London:* Allen Lane, 1973).

Raymond Williams, *Problems in Materialism and Culture* (London: Verso, 1980), 'Ideas of Nature', pp. 67-86.

Raymond Williams, *The Country and the City* (London: Chatto & Windus, 1973), ch. 12: 'Pleasing Prospects'.

Raza Moonis and Aijazuddin Ahmad, (1977) *India--A General Geography*. New Delhi: National Council of Educational Research and Training.

Richard Chorley, Stanley Schumm, and David Sugden, *Geomorphology*. Edition Methuen, 1984.

Robert G. Wetzel & Gene E. Likens, *Limnological Analyses*, 3rd ed. (Springer-Verlag, 2000).

Robert G. Wetzel, *Limnology: Lake and River Ecosystems*, 3rd ed. (Academic Press, 2001).

Ronald Amundsen. Soil Preservation and the Future of Pedology.

Royer, Dana L. and Robert A. Berner, Isabel P. Montañez, Neil J. Tabor, David J. Beerling (2004) "CO2 as a primary driver of Phanerozoic climate", in *GSA Today* July 2004.

Ryszkowski, L. (ed.). 2002. Landscape Ecology in Agroecosystems Management. CRC Press, Boca Raton, Florida, USA.

Sanderson, J. and L.D. Harris (eds.). 2000. Landscape Ecology: A Top-Down Approach. Lewis Publishers, Boca Raton, Florida, USA.

Schaefer, F.K. (1953): Exceptionalism in Geography: A

Methodological Examination, A.A.A.G., Vol. 43, pp. 226-245.

Shaviv, N. and Veizer, J. (2003) "Celestial driver of Phanerozoic climate?", in *GSA Today* July 2003, volume 13, number 7, pages 4-10.

Troll, C. 1939. Luftbildplan und ökologische Bodenforschung (Aerial photography and ecological studies of the earth). Zeitschrift der Gesellschaft für Erdkunde, Berlin: 241-298.

Turner, M.G. 1989. Landscape ecology: the effect of pattern on process. Annual Review of Ecology and Systematics 20: 171-197.

Wetson, J.W. (1957), The sociological aspects of geography, in G. Tailor (ed) Geography in the twentieth century. Methane, London.

Williams, *Problems in Materialism and Culture,* 'Social Darwinism'.

Wilson, J.B. and W.M. King. 1995. Human-mediated vegetation switches as processes in landscape ecology. Landscape Ecology 10: 191-196.

Index

D

E

F

G

Q

R

S

T

U